WEATHER WORLD

PHOTOGRAPHING THE GLOBAL SPECTACLE

WEATHER WORLD

PHOTOGRAPHING THE GLOBAL SPECTACLE

GORDON HIGGINS

IN ASSOCIATION WITH
THE MET OFFICE

WORLD LEADER IN CLIMATE CHANGE AND WEATHER FORECASTING

D&C
David and Charles

A DAVID & CHARLES BOOK
Copyright © David & Charles Limited 2007

David & Charles is an F+W Publications Inc. company
4700 East Galbraith Road
Cincinnati, OH 45236

First published in the UK in 2007
First published in the US in 2007

Text copyright © Gordon Higgins 2007

A catalogue record for this book is available from the British Library.

ISBN-13: 978-0-7153-2640-4 hardback
ISBN-10: 0-7153-2640-6 hardback

Printed in China by SNP Leefung
for David & Charles
Brunel House, Newton Abbot, Devon

Commissioning Editor: Neil Baber
Editor: Emily Pitcher
Art Editor: Marieclare Mayne
Assistant Editor: Demelza Hookway
Copy Editor: Constance Novis
Production Controller: Beverley Richardson

Visit our website at www.davidandcharles.co.uk

David & Charles books are available from all good bookshops;
alternatively you can contact our Orderline on 0870 9908222 or write
to us at FREEPOST EX2 110, D&C Direct, Newton Abbot, TQ12 4ZZ
(no stamp required UK only); US customers call 800-289-0963 and
Canadian customers call 800-840-5220.

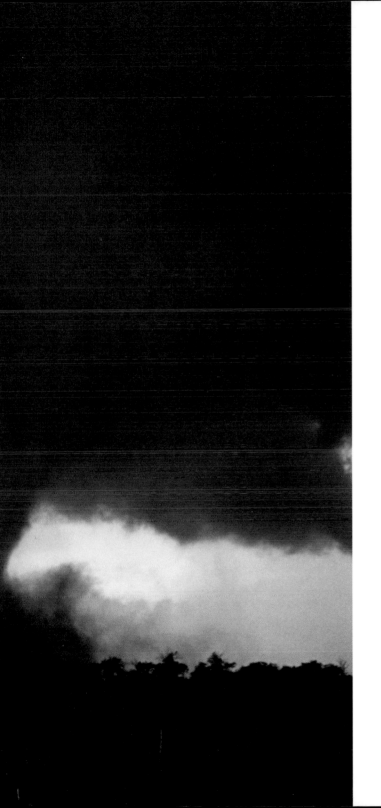

FOREWORD:

This is one of those books that has the 'wow factor'. I do not say that often, and I suspect that I am particularly hard to impress, having spent some twenty years looking at images and observations of the weather as part of my day job. It does make me want to go out and buy a better camera.

Reading through the many beautifully put together and eye-catching pages it reminded me of some of the things that made me interested in meteorology to begin with. The first is that we live on an impressive and stunningly beautiful planet, which we could all do more to take care of. The second is the sheer scale and power of the world's weather, and the influence it has on our daily lives. The third is how nature, and the weather in particular, know no boundaries: social, geographical or political. The atmosphere is a living, moving system where inter-connected weather features of many different physical sizes are born, then grow and decay, over varying timescales. Finally, I am not sure that anyone could have imagined how the technology of observation systems such as satellites would help to unlock the many secrets of our weather patterns and features and the underpinning processes that drive their evolution – and there are some stunning examples in this book.

I challenge anyone not to be impressed by what they see in these pages. I am sure it will spark your imagination, but I also hope it will encourage you to find out more about our weather and climate – take it from me, it is a fascinating topic.

Professor Paul Hardaker, CMet
Chief Executive, Royal Meteorological Society

INTRODUCTION:

Weather World celebrates and brings into focus the majesty, power, and dramatic beauty of the Earth's weather. The book takes two views of the world's weather, from above and from below. From satellites, we see the splendour of the Earth itself, setting the background to the weather we experience. Through the same perspective, we explore and expose the large-scale swirling patterns of the Earth's major weather systems, and some of the remarkable cloud patterns found across the globe. And we show examples of the beauty of the most powerful small-scale weather systems.

From below, and on a smaller scale, *Weather World* takes a look at individual clouds and cloud clusters, and some of the colourful effects they produce. We will explain why you see rainbows and other coloured patches in the sky, including halos, arcs, and iridescences. They are illustrated here in all their glory.

Against this backdrop of beautiful day-to-day weather is the certain knowledge that the Earth's climate is changing. *Weather World*'s imagery shows you some of the serious effects of climate change. Perhaps these dramatic pictures will help us realize what is happening to our planet, and then provoke a personal response to helping slow down the rate of change.

In one way or another, the weather affects us all. Wherever we live, there are aspects of weather we have to take into account. It may be dressing for the heat of summer or for the depths of a winter freeze. It may be coping with prolonged dry periods or perhaps with torrential tropical downpours. One thing is sure – weather is here to stay.

THE DRIVING FORCES OF THE WORLD'S WEATHER

The Sun is the powerhouse of the planet, providing vital heat to the Earth's surface. The moisture in the air leads to clouds, rain, and snow, providing the water for our everyday lives. And the Earth's rotation is the giant mixer that moves the heat and moisture around. These three create all the amazing weather features we experience every day.

Without the Sun, life on our planet would be impossible. We would experience incredibly low temperatures, far too cold to sustain any form of living organism. Without moisture, there would be no oceans, no lakes or rivers…and no clouds. We would be living on a barren, desert wilderness, just like that of the planet Mars. And without the rotation of the Earth, we would lose the familiar pattern of day and night. The mixing of the atmosphere would be hugely reduced and thus many of the weather features pictured in this book would not exist: no hurricanes, typhoons, tropical storms, tornadoes, and no areas of low pressure.

The Sun. Moisture. A rotating Earth. Put them together and we get weather. Without any one of these, weather simply would not exist.

> "WEATHER WORLD CELEBRATES AND BRINGS INTO FOCUS THE MAJESTY, POWER, AND DRAMATIC BEAUTY OF THE EARTH'S WEATHER"

OBSERVING AND PREDICTING WEATHER

Once, no one knew what caused the weather. Many civilizations attributed it to their gods, and sought the weather they wanted by performing rituals and offering sacrifices. It was only in the early 1800s that Admiral Frances Beaufort created the first scientific systems for measuring and recording the weather consistently. His wind scale and weather codes provided a logical basis for simple weather observations, resulting in an understanding of bigger weather systems.

Slowly, the science of meteorology evolved. Admiral Robert FitzRoy developed the barometer, used for measuring atmospheric pressure. The burgeoning telegraph network made it possible to communicate and collect information internationally. By the time Admiral Fitzroy became Head of Meteorology at the Board of Trade in 1854, he was well positioned to start producing simple weather forecasts. Now a worldwide weather observing network was taking shape.

Since those early days, methods of observing the weather have changed dramatically, the most significant innovation being the introduction of weather satellites in 1960. For the first time ever, the cloud formations that covered huge areas of the Earth were visible in one image. By studying these, meteorologists began to recognize patterns and to interpret their movements. This made weather forecasting much more accurate.

What is not so commonly understood is that weather satellites do not just measure temperature, but also humidity and wind. All these measurements are sent to meteorological centres, like the UK's Meteorological Office (known today as the Met Office).

Numerical weather prediction uses complex and sophisticated computer models of the atmosphere to analyze and predict weather on a global scale. It forms the basis of all weather forecasting today. Billions of pieces of data go into producing these forecasts. Taken from right round the globe and at many levels in the atmosphere, these observations are first analyzed to get a global picture of the world's weather at that moment. Then some of the world's most powerful super-computers work their way through billions of calculations to create today's amazingly accurate forecasts.

WEATHER PATTERNS, LARGE AND SMALL

On a global level, the effects of the Sun, moisture, and the rotating Earth create persistent patterns of wind and pressure around the planet. High pressure is the result of descending air with winds blowing clockwise at the Earth's surface; low pressure results from rising air and surface winds blowing anti-clockwise. There is a low pressure band at the equator, and there are two more bands of low pressure in the temperate latitudes towards the Arctic and the Antarctic. Pressure at the poles is generally high, as it is in the two bands close to the Tropics of Cancer and Capricorn.

Because air always flows from high to low pressure, these global pressure bands generate global wind patterns. These will change with the seasons as the northern and southern hemispheres of the tilted Earth move closer to, or further away from, the Sun.

In tropical regions, the weather doesn't change much from day to day or season to season. It is generally hot and pretty humid with most of the rainfall coming from thunderstorms. The farther away from the equator the greater seasonal variations become. But, at the poles, the weather is always cold or very cold.

The sub-tropical and temperate latitudes generate the world's most interesting and changeable weather. During the sub-tropical summer, the sun warms up large areas of the sea's surface. It is here that tropical cyclones are born. They are known as hurricanes in the Atlantic Ocean, northeast and southern Pacific Ocean, as typhoons in the northwest

"WHEREVER WE ARE, THERE ARE ALWAYS BEAUTIFUL SIGHTS TO SEE IF WE SIMPLY TAKE A LITTLE MORE TIME TO LOOK UP"

Pacific Ocean, and as severe cyclonic storms or tropical cyclones in the Indian Ocean.

In recent years, tropical cyclones have brought devastation to many parts of the world including East Africa, the Indian subcontinent, and south-east Asia, particularly the Philippines. Atlantic hurricanes have also badly affected the Caribbean Islands, Mexico, and the southern states of the USA. To clearly identify them, these storms are alternately given men's or women's names. In any one Atlantic hurricane season, from June to November, there are typically 10 tropical cyclones and six hurricanes, yet this is only 10–15 per cent of the total number of tropical cyclones worldwide. In 2005, the most active and destructive Atlantic hurricane season ever, there were 28 tropical cyclones, with 15 major hurricanes.

In temperate regions, we see frequent areas of low pressure, or lows, which are also known as depressions. In the northern hemisphere these form at the boundary where polar air moving south meets tropical air moving north. A developing depression usually has a leading warm front and a following cold front. These fronts take their names from the temperature of the air behind them. Much less intense than tropical storms, lows can still bring very wet and windy weather. More frequent and active in the winter months, these are the systems that bring bad weather to much of North America, Europe, and northern Asia.

CLIMATE CHANGE

Over the last 30 years or so, questions have been raised worldwide about whether or not the Earth's climate is changing. The key factor was, and remains, the increasing release into the atmosphere of greenhouse gases, mainly carbon dioxide from the burning of wood, and such fossil fuels as oil, gas, and coal. These gases form an insulating blanket around the Earth, reducing the usual loss of heat into space, and resulting in the noticeable warming of both the atmosphere and the Earth.

Two questions must be asked. Is the climate changing? And, if so, is mankind responsible for that change?

Staff at the UK Met Office started working on climate research back in the early 1970s and in early 1989, they consolidated that research within the Hadley Centre for Climate Research. In recognition of the global nature of potential climate change problems, the World Meteorological Organization and the United Nations Environmental Programme set up the Intergovernmental Panel on Climate Change (IPCC) in Autumn 1988 to explore the issues surrounding climate change and to come up with 'the answers'. Initially, scientists used huge computer models designed to include all the known factors that could affect the atmosphere. More recently, it became clear that these models would have to include links with the world's oceans. Thus, very detailed and complex 'coupled' ocean-atmosphere models were born.

So what is their answer? In early 2007, scientists within the IPCC programme updated their earlier conclusions stating even more emphatically that our climate is changing – the burning of fossil fuels in particular is releasing huge amounts of carbon dioxide into the atmosphere, leading to a warming of our planet. Mankind is responsible for changing the climate.

What evidence is there for their conclusion? In the twentieth century, records show a slowly rising trend in global temperature of about 0.5°C, and in the northern hemisphere temperature has increased more than during any other period in the last 1000 years. What is more, globally the 1990s was the warmest decade of the millennium, and the first years of the twenty-first century are continuing that trend. The majority of glaciers are melting and retreating. The Arctic ice cap is shrinking and thinning. There is a consensus that in the future we will experience hotter summers and more widespread drought and there is continuing debate about other ways in which our weather might change. Climate scientists are working to answer crucial questions: will we see increasing numbers of gales? How will rainfall patterns be affected? What will happen to the intensity and frequency of tropical cyclones?

NATURE'S RICH PAGEANT
And so we come full circle, back to the sheer beauty of our day-to-day weather. Perhaps you will be amazed by the awesome power of hurricanes, by cumulonimbus clouds with their dramatic displays of thunder and lightning, or by tornadoes twisting their destructive paths. Or perhaps you prefer the fleeting colours of a rainbow, the spectacular effects of sunset over mountains, or the eerie flickering beauty of the aurora borealis.

These natural features are there for us to see, in some form or another, every day. True, some parts of the Earth experience more dramatic weather events than others. Yet wherever we are, there are always beautiful sights to see if we simply take a little more time to look up.

I have enjoyed watching the weather for more than 50 years, yet I never tire of seeing nature's rich pageant unfolding. Even today, as I write, the morning's rain is dying away and the cloud is starting to clear. A frontal system is crossing the UK, with a sharp change from rain and cloud to clear skies and sunshine; from warm damp air to colder, crisper conditions.

I trust that through the images in this book, you will join me in celebrating the true beauty and wonder of our *Weather World*.

Gordon Higgins, 2007

PART ONE: WEATHER FROM ABOVE

Weather World is split into two perspectives: in the first we look down at the weather from the sky, mainly from satellites orbiting our planet. Some of the images are actually taken by astronauts looking down as they fly around the Earth. This is weather on a grand scale.

CHAPTER ONE:
WORLD MAPS AND WIDER VIEWS

We begin our exploration of the world's weather from the glorious vantage point of satellites in space. We look at pictures showing whole discs of the Earth, taken from geostationary satellites, those with orbits perfectly synchronized with the rotation of the Earth, always looking down at the same spot on the planet's surface. This is where we can, in one snapshot, get an instant overview of global weather patterns.

We then start to focus on an interesting variety of regional weather features, those that occur in particular areas on specific continents or oceans. In the North Atlantic these are low pressure frontal systems. In the Mediterranean, we see clear skies alongside desert, clouds and snow cover. And in North America, clear skies reveal the snow-covered vastness of the Great Lakes.

Many of the images shown are from orbiting satellites, those that travel around the Earth several times a day, usually going over the poles; they look down on different slices of the Earth each time they go round. Whether black and white or full colour, these global and regional images give us incomparable overviews of our *Weather World*.

GLOBAL SYSTEMS

THE BLUE MARBLE

This pair of images of planet Earth gives us the widest view we can get of where we live. They are part of a series known as the Blue Marble. In these pictures, scientists have combined images and data from several instruments on a number of satellites, parts of the scientific discipline of remote sensing. The pictures include information about the height and shape of land masses, the type of vegetation that covers the land, the nature of the ocean (shallow or deep) typified by its colour, ice cover in the polar regions, and finally, the cloud formations that give us our weather.

The image on the left focuses on the Indian sub-continent with East Africa, the Horn of Africa, and the Middle East clearly shown on its left. Huge swathes of blue give the proof of just how much of our planet is covered by oceans.

On the right, the main focus is North and Central America and the Caribbean. Again, we can see the vastness of the Pacific Ocean that takes up more than half the image.

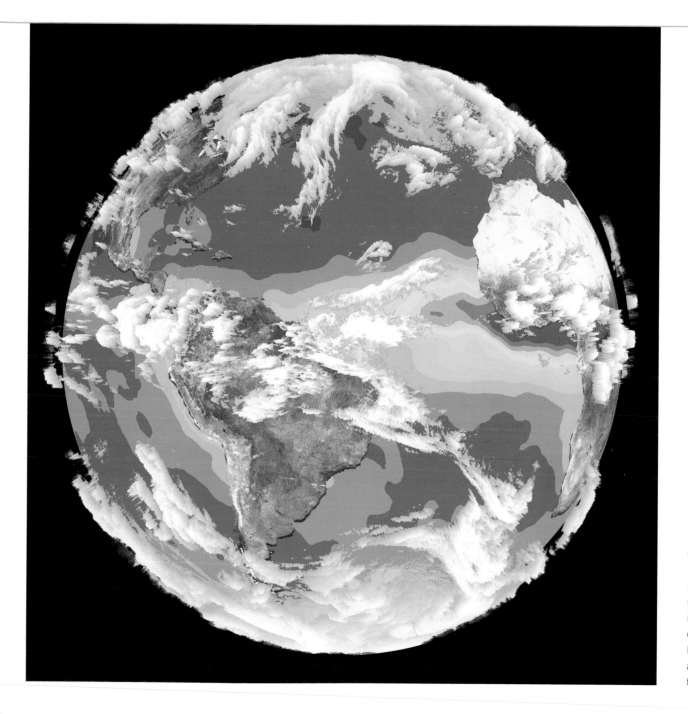

◄ 3D ATLANTIC DISC

Satellite data and images from four different sources contributed to this 3D picture of half the Earth. The true-colour land and vegetation type come from one source, and this was augmented by another higher resolution data stream that picks out even the detail of fires (red dots) across Central and South America, and Africa. The data for the oceans comes from yet a third satellite source, while the cloud images come from a composite of four different geostationary satellites. The overall purpose of the image is to give us as clear a picture as possible of what the Earth is really like.

◄◄ THE FLORIDA TURTLE

This image gives us an unusual view of the south-eastern states of America, particularly Florida. Looking rather like a turtle sticking its head out of its shell, the state and those to the north represent a huge area of the USA basking in a cloudless sky. Just a few speckled areas of cloud appear over the Atlantic and the Gulf of Mexico. The shallow waters off the Florida Keys and The Bahamas show up turquoise against the dark blue of the deeper ocean.

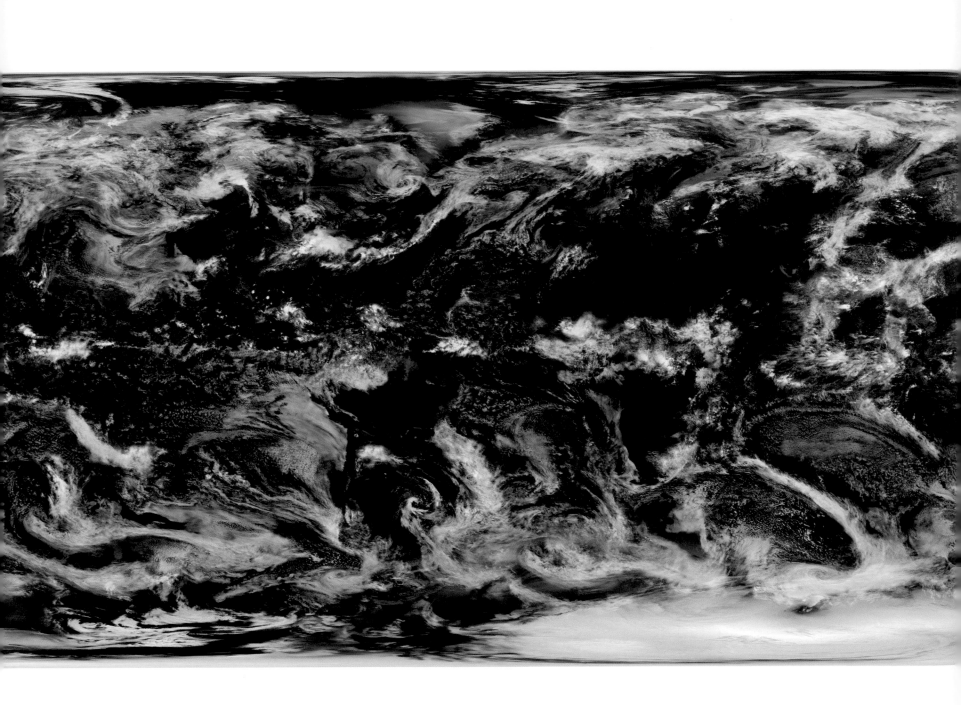

◄ BLUE MARBLE CLOUD STRIP

In this unusual perspective of the Earth, we see the clouds from the Blue Marble series unwrapped from the globe and presented in black and white. It is worth noting that this unwrapping process makes areas of cloud appear to increase disproportionately in size as we move from the equator towards the poles. Despite this, it is clear that there are generally fewer clouds over the equator and the sub-tropics; many more clouds are present in the temperate latitudes, particularly in the southern regions. Weather forecasters carefully analyze how all these cloud and weather systems develop, move, and disappear.

COLOURED BLUE MARBLE STRIP ►

This image overlays the land and sea with the clouds from the black and white picture (left). The primary source for this composite image is the instrumentation on board NASA's Terra satellite, flying 700 km (435 miles) above the Earth. Among other data, it provides us with the cloud imagery. Land and coastal ocean portions of the picture came from surface observations; ocean and ice data are added in. It shows our world wreathed by the beauty of the patterns and features of its cloud systems.

FORECASTING TECHNIQUES

METEOSAT AFRICAN DISCS

Images like these two captured by Meteosat, the European Space Agency's geostationary satellite, provide weather forecasters with a wonderful tool for analyzing our world's weather. Together with images from sister satellites that look down on the rest of the globe, it is possible to get a complete picture of what is going on weather-wise. By using these images with charts based on computer analyses of surface and upper-air weather observations, the forecaster is better equipped than ever before to provide accurate predictions.

◄ In the smaller image, much of southern Africa is completely cloud free, as is a large part of South America (to the left of the picture). A loose broken line of cloud running almost east/west marks the inter-tropical convergence zone that roughly marks the boundary between the Earth's two hemispheres; even the Sahara region has some cloud cover.

In the larger image, southern Africa is now largely covered with cloud: the rainy season has arrived. To the north, the Sahara is almost free of cloud. And further north, over the Atlantic Ocean, we can see the familiar swirls of the low pressure systems that regularly affect Europe. ➤

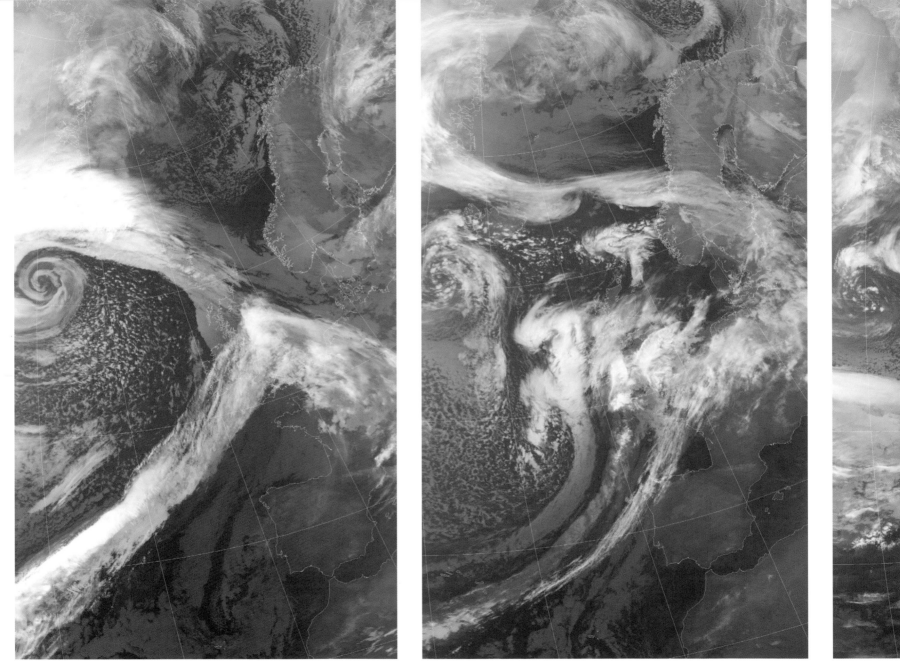

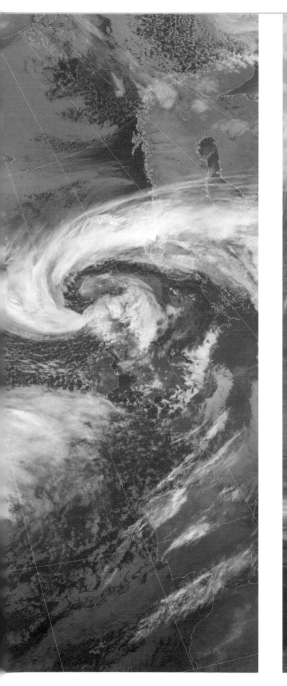

UK STORMS OF OCTOBER 2000

In England and Wales, the autumn of 2000 was the wettest since records began in 1766. In October 2000, total rainfall for England and Wales was more than twice the average, an exceptional result. Moreover, the rain was accompanied by ferocious winds at the end of the month, and produced significant flooding. This sequence of images from the 27 to 30 October captures the low pressure storms and their frontal cloud bands as they sweep across the UK.

The first image shows the marked swirl of the main low pressure area near Iceland with speckled shower clouds to the south and fronts stretching across the UK. The real drama starts on 28 October (centre left) when the first of a series of so-called daughter depressions brought torrential rain, gale force winds, and even a tornado in Bognor Regis. Although 29 October started dry and bright for many (centre right), forecasters were already watching to the west where the next system was spinning up to bring more very heavy rain. The previous day's fronts are now wrapped around northern Britain. And finally (right), when the last storm tore into the UK with winds gusting up to 205km/h (128mph) and another 2.5–7.5cm (1–3in) of rain, the floods really set in.

EUROPEAN HEATWAVE OF AUGUST 2003

This pair of images taken at the same time on the same day, 1257 UTC on 5 August, shows two different views of a very hot Europe. There is a visual image, like a normal photograph, and an infrared image, which measures the temperature of the clouds and the surface of the Earth.

In the visual image on the left, the ground shows as grey, and the sea as shades of dark grey or black. It is possible to get a feel for the nature of the terrain in the mountainous regions, the Alps and Pyrenees showing well with a mix of cloud and snow on the peaks. Elsewhere clouds can be seen as white areas. Some are large and some very small, almost like spots.

The infrared (temperature) image shows things dramatically differently. The record-breaking heat of the land shows up very black while the much cooler sea is grey, which is a reversal of the visual image. Cold cloud tops show bright white. This difference between images can be of great help to forecasters.

The heatwave was a real record breaker. Temperatures reached between 40° and 50°C (104°–122°F) in many countries, and tragically, the heatwave was responsible for between 30,000 and 50,000 deaths across Europe, and significant damage to crops.

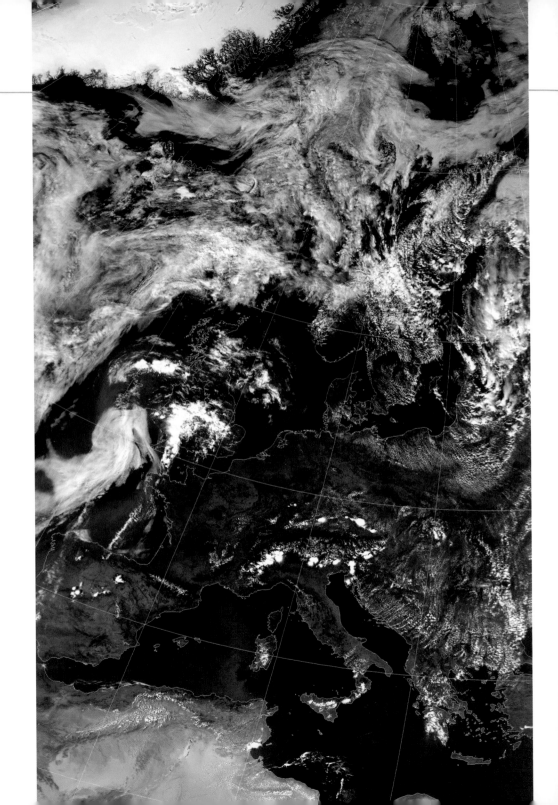

◄ DEPRESSION OVER CANADA

Low pressure systems, or depressions, mean bad weather. The centre of this summer depression is sitting right over Hudson's Bay, in Ontario, Canada, with the swirling cloud from its associated frontal systems draped across parts of Manitoba (to the left), Ontario (bottom centre), and Quebec (right). It is the spiralling cold front that is most prominent, with just a short part of the warm front visible towards the top right. An area of shower cloud can be seen tucking in behind the cold front in the bottom left quadrant of the picture.

DEPRESSION OFF ICELAND ➤

This classically formed system of low pressure in the Denmark Straits (between Greenland and Iceland) clearly shows the classic northern hemisphere anti-clockwise flow of winds and frontal cloud into its centre. Captured in September 2003, its shape resembles the spirals of a snail's shell. Air moving from high to low pressure sweeps into the depression, rises into the atmosphere, and forms the characteristic frontal cloud structures seen here. The north-east part of Iceland is clearly visible towards the top right of the picture.

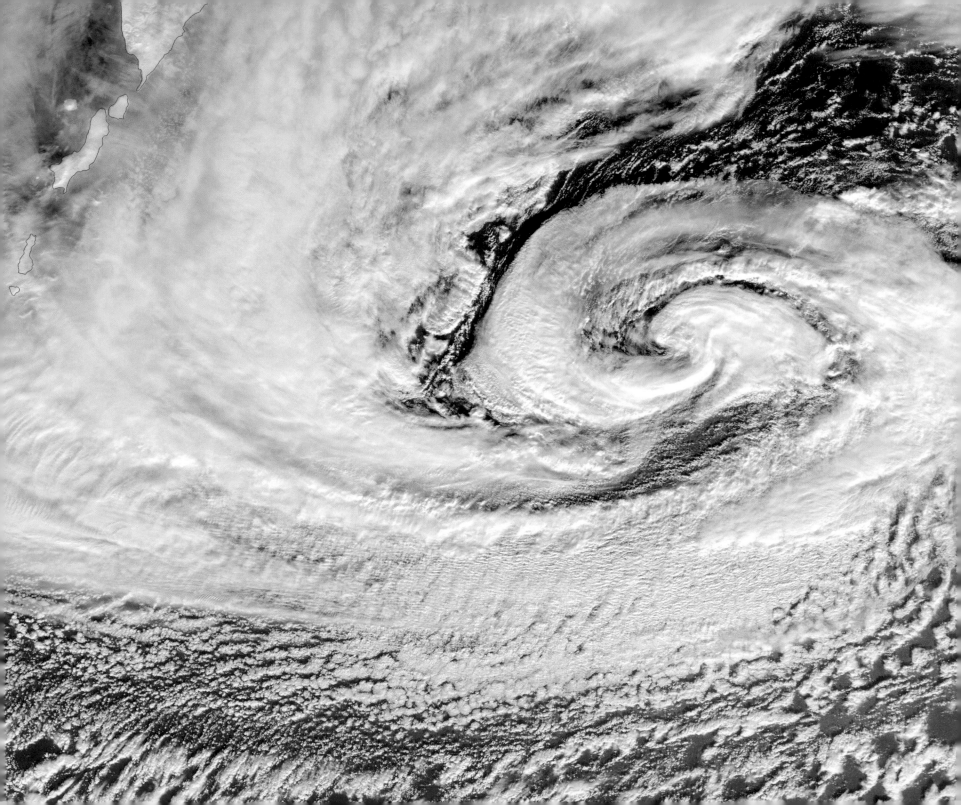

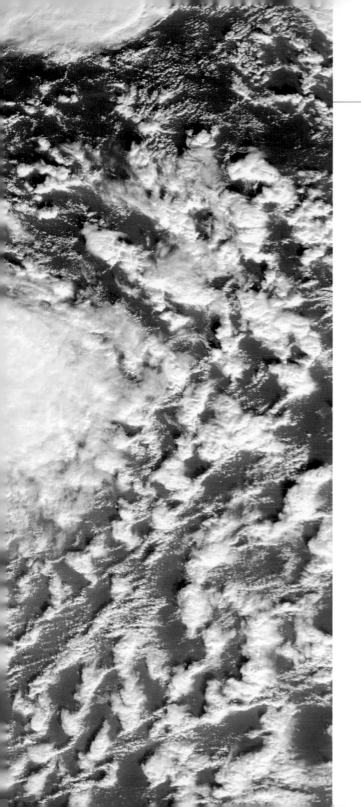

LOW OVER THE NORTHWEST PACIFIC

This impressive low pressure system was captured by NASA's Terra spacecraft in December 2002. The low is lying just south of the Russian Kamchatka Peninsula and the Aleutian Islands; the snow-covered tip of the peninsula can just be seen in the top left-hand corner of the image. This is a well matured depression, its frontal cloud bands wrapping right into the centre of the low. On the southern and eastern flanks of the low, the dense frontal cloud layers give way to beautiful broken lines and swirls of shower clouds.

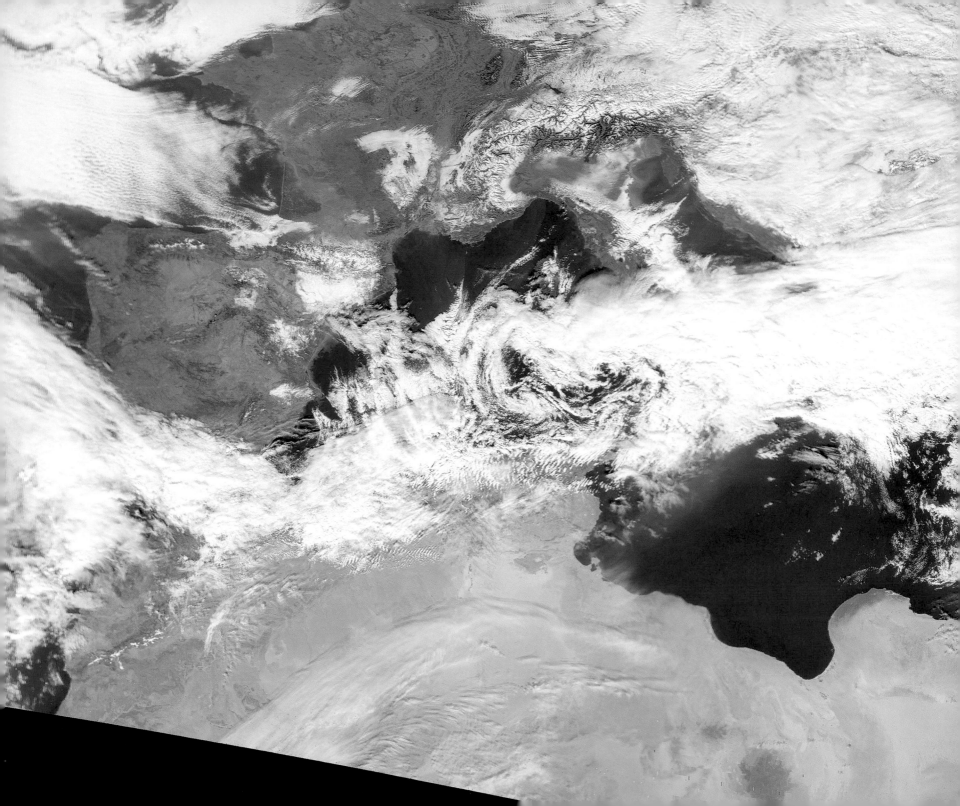

CONTINENTAL SCALE WEATHER PATTERNS

WINTER STORM ACROSS EUROPE AND THE MEDITERRANEAN

A severe storm swept across much of Europe on 16 December 2001. It brought a combination of heavy snowfall, strong winds, and bitterly cold temperatures. The effect was to bring many transport systems to a halt and to cut off power to hundreds of thousands of people. This true-colour image, captured the day after the worst of the weather, shows the heavy cloud from the storm with snow over the Alps, northern Italy, the Pyrenees, and southern France. It is also possible to see wind-borne dust sweeping out into the Mediterranean from the North African coast near the Libya–Tunisia border.

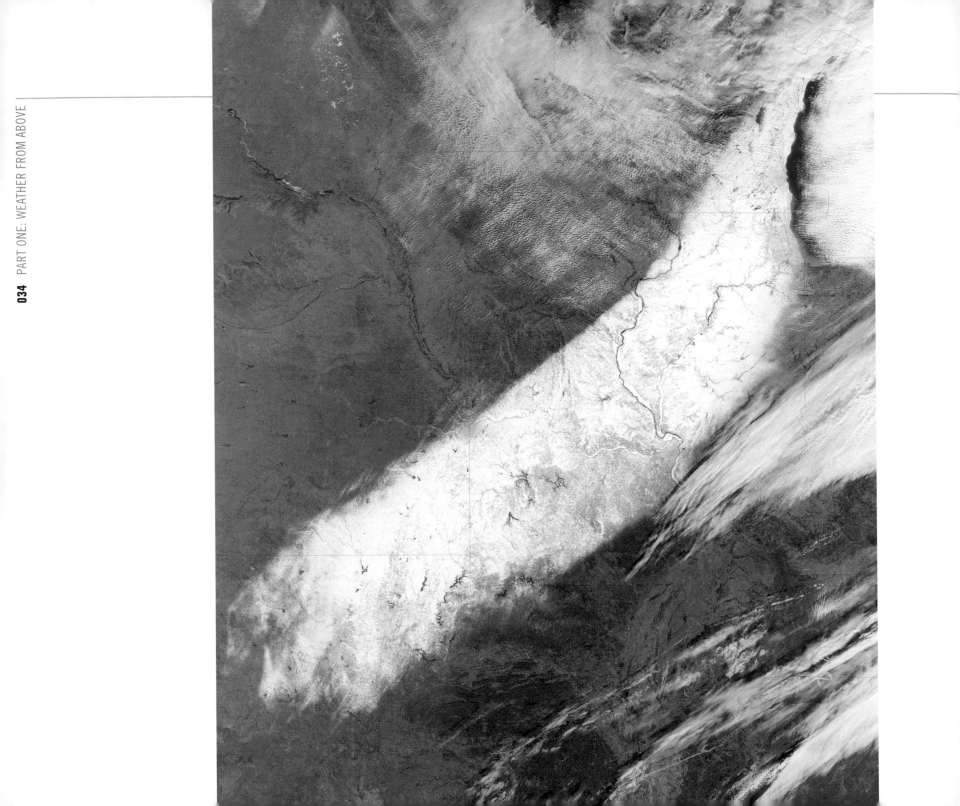

◄ HEAVY SNOW OVER THE AMERICAN MIDWEST

A severe winter storm hammered the Midwestern USA on 1 December 2006. It caused widespread disruption: the icing of roads; cancellation of airline flights; damage to trees and branches; cutting off of electricity supplies to more than two million properties; and temporarily shutting down part of a major motorway. Several deaths were linked to the storm. What we see here is the aftermath, a giant white smear of snow against the tan background. By the time of this image, the system had moved on into Canada leaving the Midwest almost free of cloud, but with many people struggling to restore life to normal.

FOG OVER LONDON ►

Thick fog surrounded London on the afternoon of 20 December 2006 when NASA's Aqua satellite captured this image. The thickest fog is west of London, the second white 'blob' from the centre-right edge, north of the Channel coast. London's primary airport, Heathrow, is located under the fog bank. As a result, many flights out of the airport were cancelled, leaving up to 40,000 travellers stranded. There are also some delicate wave clouds over north-east Ireland and some lovely cloud streets over Biscay (bottom centre).

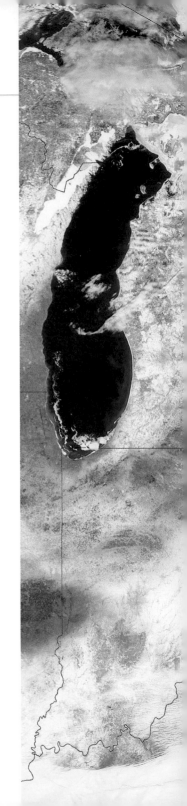

◄ FOG PLUMES OVER THE GREAT LAKES

Unseasonably warm, moist air is the driver of the effect seen in this image of Lakes Michigan and Huron in April 2002. The air was around 20°C (68°F) while the water temperature was below 5°C (41°F); cooling by the water surface produced fog, which has grown into fog plumes (in the middle of the black lakes) as the air is pushed forward on the wind. Once the fog meets the land, we see some fascinating patterns forming as the fog reflects off the coast and bounces back, rippling like a shock wave. This is particularly apparent on the north-west coast of Lake Michigan and in one pocket of western Lake Huron. (Land shows as green, ice and snow as red.)

PRESIDENT'S DAY SNOW STORM 2003 ►

Affecting a huge part of the mid-Atlantic and northeastern USA and Canada, the President's Day storm of 14–19 February 2003 was a record-breaking blizzard. This picture was captured two days after the storm had moved on, leaving 38–76cm (15–30in) of snow over a very wide area and many cities brought to a standstill. The severity of the event was caused by a classic combination of weather systems: a developing depression feeding on the warm, moist southerly flow of Atlantic air, and an area of high pressure over eastern Canada that fed cold northerly air into the heart of the low, to devastating effect.

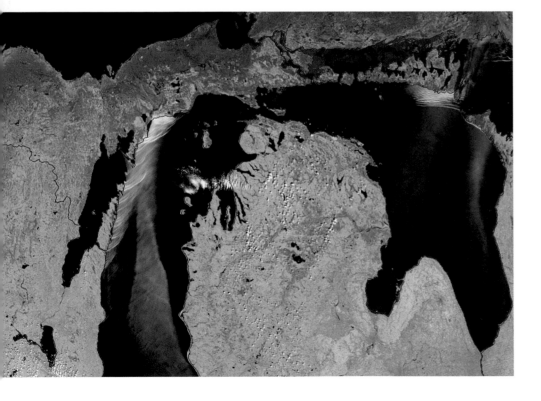

◄ CLOUDLESS EASTERN US AND CANADA

Looking like a map in an atlas, this image marks a rare occurrence over North America – such a large area of the continent free of cloud. The clear skies were associated with a large area of high pressure over the eastern central US, frontal cloud having swung through the region and out to sea.

POLLUTION BLOWS OVER THE CHINA SEA ►

The atmosphere is something we share with our neighbours, near and far, and not always to our advantage. This picture shows cloud and a thick layer of pollution flowing out of eastern China, across the China Sea, the Korean Peninsula, and towards Japan. China relies extensively on its massive resources of coal to fuel its power stations. While this generates cheap electricity, it sadly also generates copious amounts of polluting emissions. Add in the output from an ever-growing number of cars and the problem escalates. What is good for one nation is not always good for another.

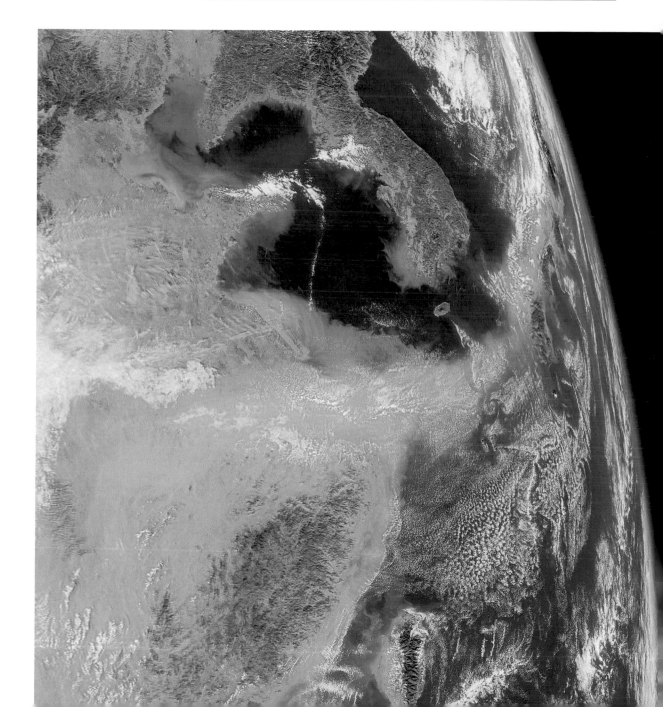

NORTH AMERICA –
THE BIG PICTURE

This striking cloudscape of North America illustrates a range of things to see from space. Off the west coast, intricate cloud patterns back up against the land, not able to come ashore. In the northeast, thicker white cloud indicates areas of low pressure. Over the midwest, some small but chunky areas of white cloud suggest thunderstorm activity, as do the large areas of bright white cloud in the Gulf of Mexico. There are also some smoke plumes (thin grey streaks) drifting from west to east over Manitoba and some smaller white ones just south from the western corner of Hudson's Bay.

"SATELLITE IMAGES REVEAL
AMAZING DETAILS OF THE
COMPLEX PROCESSES HAPPENING
ABOVE OUR HEADS"

CHAPTER TWO:
CLOUD FORMATIONS

For a weather forecaster, the great joy of satellite pictures is that they – and we – can get a clear picture of clouds on a large scale. From the ground, we see only a limited patch of the sky, just a few miles round. From a satellite, we get an eye-in-the-sky view over many hundreds of miles. This allows us to see patterns in the clouds that are hidden from our earth-bound perspective.

Over large parts of the globe, it is the huge areas of high and low pressure that dominate the weather. Low pressure depressions usually have weather fronts of cloud and rain or snow associated with them, characterized by the huge patterns of swirling cloud we saw in Chapter One.

At the same time, much of the Earth will be free of cloud, or covered with cloud patterns that hold no great threat of bad weather. Such patterns can be beautiful in their own right, and often reveal amazing details of the complex processes happening above our heads. In this chapter, we explore fascinating satellite images of the cloud formations that regularly take shape in the Earth's skies.

CLOUD
FORMATIONS

◄ CLOUD STREETS IN MOZAMBIQUE CHANNEL

This true-colour image captured by satellite on 16 August 2002 shows a beautiful pattern of cloud streets in the Mozambique Channel between Madagascar (east) and southeast Africa. These lines of cloud, forming parallel to the wind, are created when continual streams of air are warmed as they are driven on a steady breeze over a warmer surface. The combined effect of upwards and forwards motion causes the air to roll and spin, like a vortex on its side. Where the rising air in adjacent vortices combine, cloud is produced; where the air descends, the sky remains clear.

CLOUD ARCS, WESTERN PACIFIC ►

Convection – direct heating of the Earth's surface by the Sun – is the driving force behind this interesting cloud pattern. The larger cloud areas inside the arcs are more vigorous than the smaller ones. Air within them rises, cooling as it does so; it then spreads out and sinks back towards the sea surface, again spreading out. The particular combination of wind, temperature and humidity has allowed the descending air to push the smaller cumulus clouds into these almost semi-circular arcs.

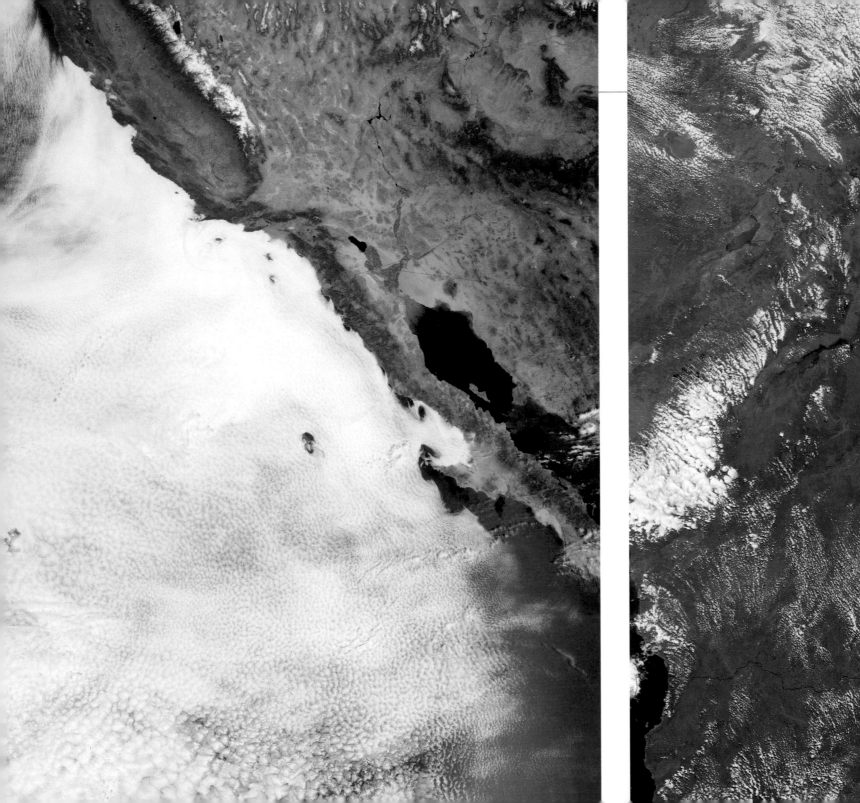

◄◄ STRATOCUMULUS OFF CALIFORNIA

Sheets of stratocumulus clouds occur readily over the Earth's oceans. They play an important part in the Earth's energy budget because they reflect a large amount of solar energy back into space. This huge cloud bank is packed against the Californian coast, anchored there by the cool ocean beneath. It is unable to spread inland because much higher land temperatures break it up. Taken in March 2002, it is possible to see the affects of spring on the San Joaquin Valley (green area, top left), while snow still covers parts of the Sierra Nevadas.

◄ TANZANIA CONTRAILS

This image of East Africa takes in Northern Mozambique and Tanzania up to the border with Kenya. Out over the Indian Ocean, and well above it, there are areas of cirrus cloud and decaying condensation trails from jet aircraft. Towards the top, the Serengeti Plain peeps out from behind a large striped cloud, possibly cloud streets. Inland (extreme bottom left), Lake Malawi is just visible, marking another national border.

◄ CONTRAILS

This wonderful web of condensation trails (commonly known as contrails) between southeastern England and continental Europe demonstrates one of the problems caused by jet travel. The familiar lines of ice crystals that form in the wake of a jet aircraft occur high up in the atmosphere. Initially quite narrow, they slowly spread out into what looks like huge sheets of cirrus cloud (see Chapter Seven). Increasing air traffic means increasing areas of contrails which contribute to further warming of the planet.

CONTRAILS OVER BRITTANY ➤

This shot shows a similar pattern of contrails as aircraft fly out of airfields in southeastern England and France, mainly London Heathrow and Paris Charles de Gaulle. Many of these are already well spread out, the planes having passed on their way some time before the image was captured.

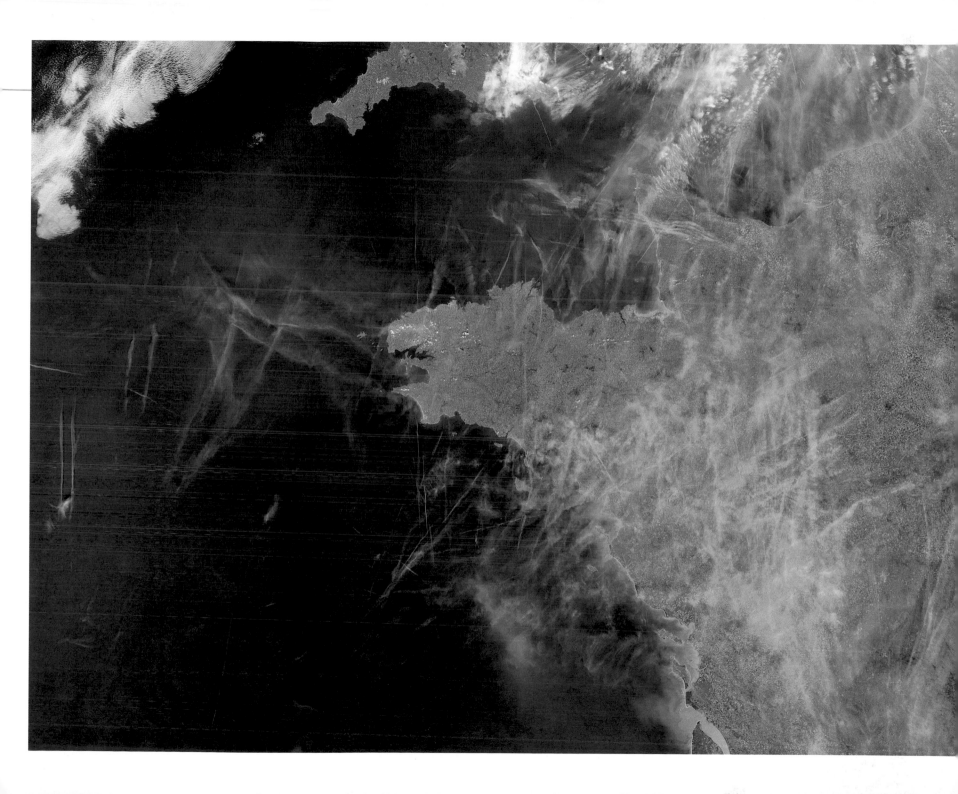

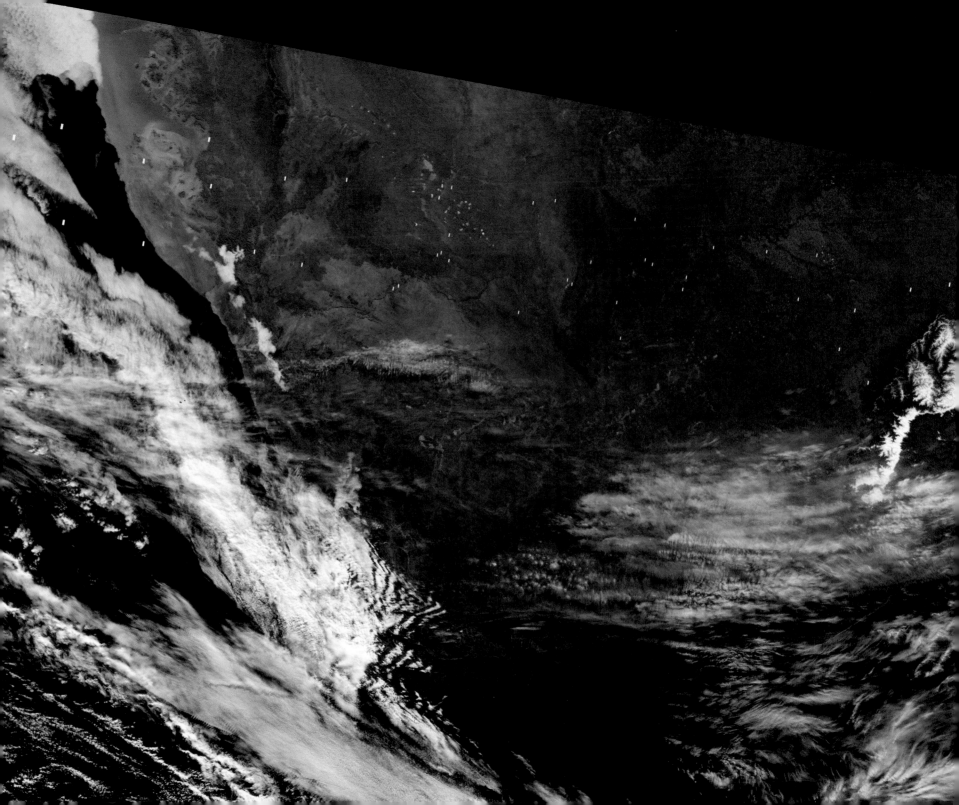

◄◄ CONVECTIVE SNOWSTORM IN LESOTHO

Lesotho is a small state in southern Africa. You may not think of snow occurring in Africa, but, as this picture shows, massive convective cumulus and cumulonimbus clouds do occur. Over Lesotho's Drakensburg Mountains, this winter storm dumped several feet of snow, causing at least 22 deaths and turning parts of the country into a disaster area. To get an idea of the scale of the storm, note the coasts of southern Africa on the left and right of the picture.

◄ THUNDERSTORMS OVER BRAZIL

This amazing picture was taken by an astronaut on the space shuttle back in February 1984. It shows some well developed thunderstorms near the Panama River in southern Brazil. You can get a feel for the powerful, boiling nature of the convection that creates these storms from the shape of the clouds near the ground (centre). The image also shows the anvil-shaped tops characteristic of thunderstorm clouds, which stretch out in all directions.

◄ CLOUD STREETS ALONG THE ALASKA PENINSULA

In this image we have a cold version of cloud streets to compare with the image off Mozambique (see page 44). The process is just the same; cold air flows steadily over the warmer sea, is warmed from below and pushed forwards. The succession of rotating streams of air produces alternate lines, or streets, of cloud and clear sky. And what intricate patterns they weave as they flow out over the Gulf of Alaska.

FRESHLY RAKED CLOUD STREETS IN THE BERING SEA ➤

Rows of small cumulus clouds stream out from the edge of the sea ice over the Bering Sea resembling the freshly raked stones in a Japanese garden. Hidden within the ice sheet, but just out of shot at the top of the picture, is St Matthew Island, the most remote and distant part of Alaska some 320km (200 miles) from the coast. It is the island that is responsible for creating the flaked and cracked ice walls that run from the top centre down to the coast on the right.

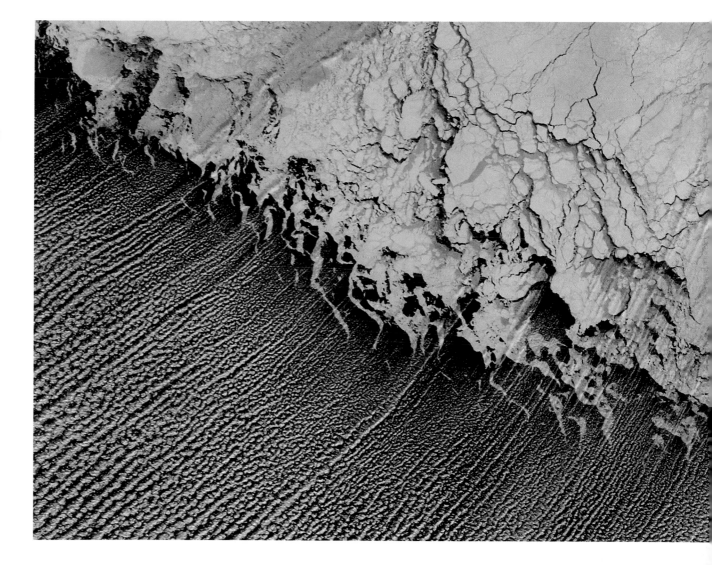

◄ TEXAS HOLE-PUNCH CLOUDS

29 January 2007 brought the unusual sight
of holes punched out of the clouds. This
satellite image shows round holes, elongated
holes, tracks, and spots, all cut into a sheet
of stratocumulus cloud. The blanket of cloud
consisted of super-cooled water droplets, which
are droplets that remain liquid even though
their temperature is well below freezing point.
As aircraft from Dallas-Fort Worth airport flew
through and over the cloud, exhaust particles
came into contact with the super-cooled
droplets, which froze instantly. The larger ice
crystals fell out of the cloud blanket leaving
behind the tiniest ice particles, and so the holes
were formed.

OPEN CLOUD CELLS OVER THE BAHAMAS ►

Open-cell cloud formation is a regular
occurrence behind low-pressure systems in
the mid-latitudes. In the northern hemisphere
a low-pressure system, with its anti-clockwise
wind flow, will draw in cold air from the north on
its rearward side and warm air from southern
latitudes on its forward side. This medium-scale
flow of air is called advection, and when cold
advection occurs over warmer waters, open
cloud cells often result. This example from
February of 2002 captures open-cell cumulus
clouds, like thick woollen strands, as the cold air
passes over the warmer Caribbean waters.

◄ HEXAGONAL CLOUD CELLS IN THE SOUTH ATLANTIC

These remarkable clouds, with their sharp edges and almost hexagonal shapes, look more like giant plates of ice than cumulus clouds. Formed by vigorous convection, they demonstrate a phenomenon called closed Rayleigh-Binard convection cells, named after the scientists who first described and explained them. They form when air rises rapidly, either from strong surface heating or, as in this case, from cooling at the top of an atmospheric layer. The closed cloud cell forms in the rising air while no cloud forms where colder air sinks. The unusual shape is simply a feature of the physical process.

CURIOUS CLOUDS IN THE EASTERN PACIFIC ►

This remarkable set of three distinctly different cloud patterns, in close proximity, was captured by the NASA Terra satellite on 18 December 2002. Closed Rayleigh-Binard convective cells in the top line have their typical angular appearance, especially in the west (left). The central line is actinoform convective cells. Taking their name from the Greek word for ray, they look like a cross between a spinal column of vertebrae and a feather boa. The line of closed convective cells at the bottom resembles an explosion of balls of cotton wool.

◄ WAVE CLOUDS, IRELAND

Large, high obstacles in a wind flow can produce interesting cloud formations. This true-colour image from December 2003 shows a wonderful set of rippling clouds stretching across the British Isles, from Scotland to southern Ireland. They have been created by the Scottish and Irish mountains. As air strikes the mountain barriers, it starts to move like a wave. As the air in the wave rises and cools, it forms a cloud; as it falls and warms, any cloud disperses. Thus the atmosphere produces a rapid succession of cloudy and clear bands, like ripples on the seashore.

GRAVITY-WAVE CLOUDS OFF AUSTRALIA ►

Two spectacular wave patterns ripple across the Indian Ocean in this image, one in the sky and one in the sea. In the upper left-hand side of the picture, we see open convective cells, but also several overlapping banded arcs of cloud. These are called gravity-wave clouds and they form when a uniform layer of air flows over a barrier like an island or mountain. In this case, it is the disturbance of different strata of the atmosphere that creates the wave pattern. Gravity waves can be carried down into the ocean too, hence the rippled effects in the sea surface on the right of the image.

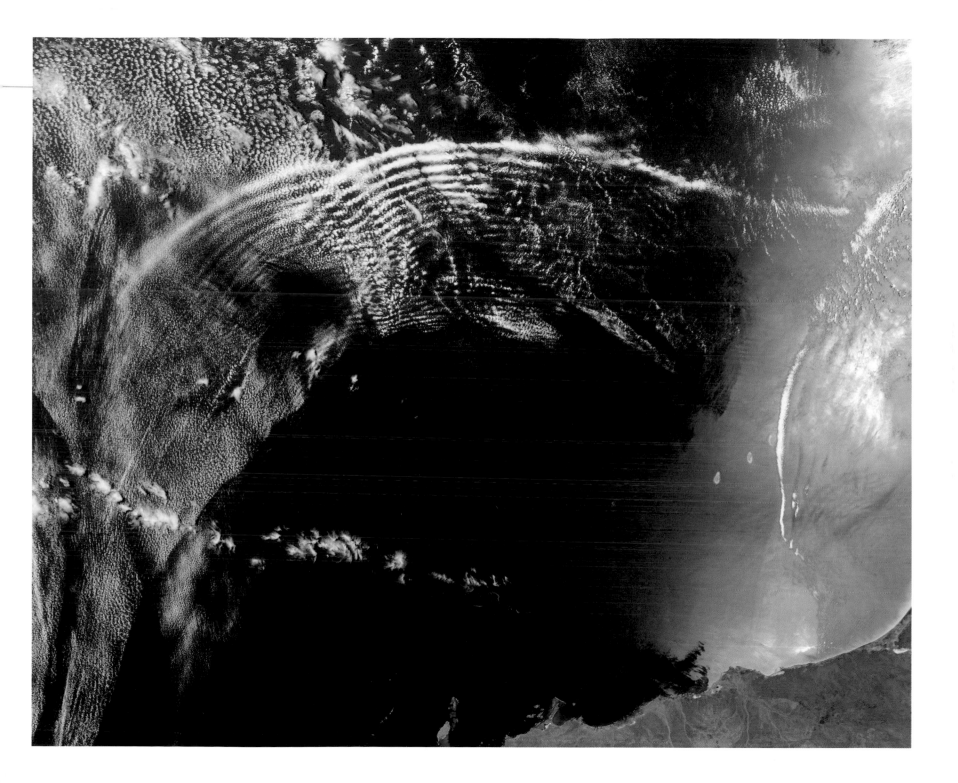

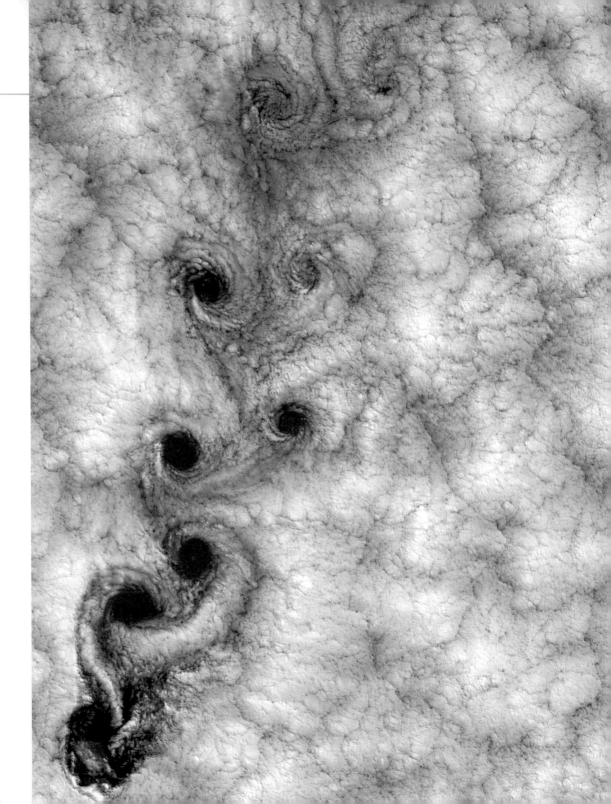

ROBINSON CRUSOE'S VON KARMAN VORTEX STREET ➤

Significant obstacles in the atmospheric wind flow can cause very large turbulent eddies. When clouds are present, the revealed eddies can be quite spectacular. This series of cloud swirls or vortices is created by wind flow past Alejandro Selkirk Island – over 160km (100 miles) west of Robinson Crusoe Island – in the Juan Fernández Islands off the Chilean Coast. The island is about 1.5km (a little less than 1 mile) in diameter and rises 1.6km (1 mile) into a layer of stratocumulus cloud. A steady equatorial wind produces clockwise vortices off the island's eastern edge and anti-clockwise flow of the western edge. The vortices grow as they travel hundreds of kilometres downwind.

SWIRLS OF LACE ➤➤

This pair of images of von Karman vortex streets reveals a much more delicate pattern than in the previous picture. These lace-like swirls of cloud are caused by the 1050 metre high Pacific island peak of Socorro, Mexico, at work in a much thinner cloud layer. The smaller-scale line of eddies to the east (right) of the right-hand image is from the tiny sister island of Isla San Benedicto. The images each show an area about 350 km (218 miles) long giving a clear feel for the scale of these amazing patterns.

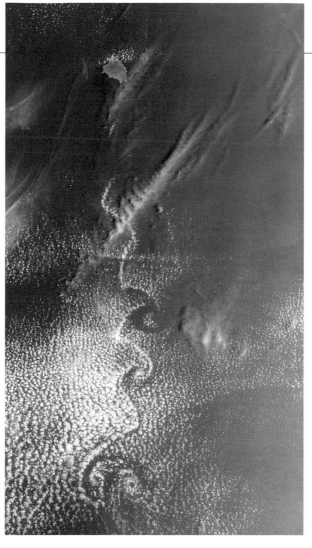

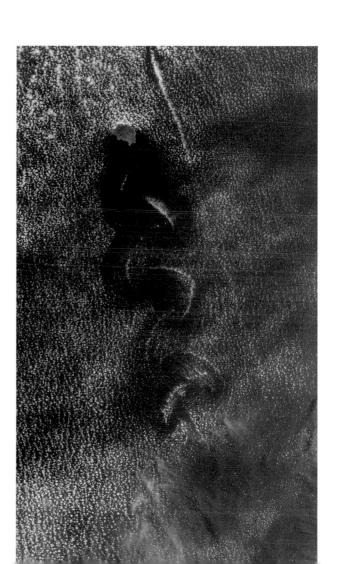

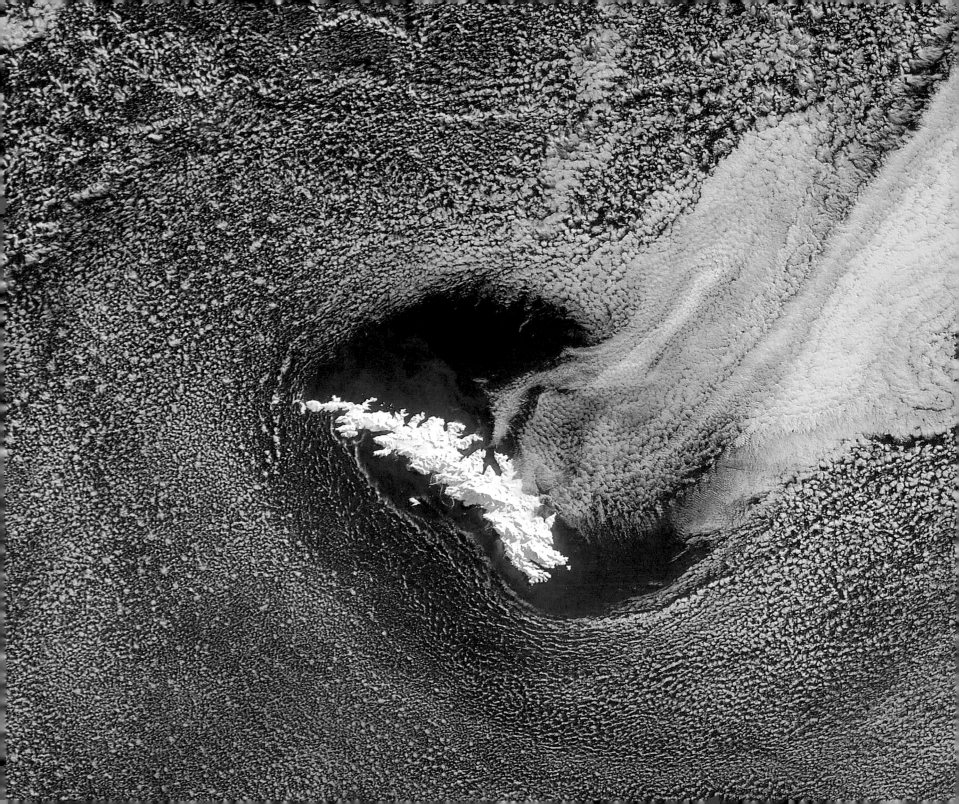

◀ AN AMAZING VIEW OF SOUTH GEORGIA ISLAND

South Georgia Island lies in the Southern Atlantic Ocean roughly 2,750km (1700 miles) due east of Argentina's southernmost tip. About 160km (100 miles) long, the island is entirely covered by snow and ice in this striking true-colour image acquired by NASA's Terra satellite on 8 August 2002. The island appears to be creating a wonderful wake of thick stratocumulus clouds flowing away from its northeastern shore, while shallow cumulus clouds distantly surround the rest of the island. There are strands of open-cell clouds in the upper left-hand side of the picture.

HIGH ISLANDS MAKE WAKES ➤

The eleven volcanic Sandwich Islands lie in a chain in the Southern Atlantic about 800km (500 miles) east of South Georgia Island. Their high peaks are creating distinct wakes in the thin stratocumulus cloud layer, like a picture of ships sailing on a still lake. The taller the peak, the bigger the wake.

CLOUD WAKE VORTICES, KURIL ISLANDS, RUSSIA ➤

The Kamchatka Peninsula is located in the far north-east of Russia, with the Kuril Islands extending like a chain south-west almost to Japan. In this image from July 2004, the peninsula is almost completely covered in cloud, just the eastern coastline peeping through. To the south, the Kuril Islands break up the cloud sheet and produce some lovely wake effects, revealing the green mountain tops of the islands beneath.

ODD FISH-SHAPED CLOUD, ROSS SEA, ANTARCTICA ➤➤

Is it a snake? Or a fish? It is neither. It is simply an amazing example of cloud pictures in the sky. The low-level grey cloud is an area of stratocumulus that has somehow formed this fish-like shape. The white eye of the fish is a large convective cloud, cumulus or cumulonimbus, which has punched its way through the low cloud and risen much higher into the atmosphere. On cooling, the air has sunk producing a dark ring of clear air around the cloud, forming the iris of the eye.

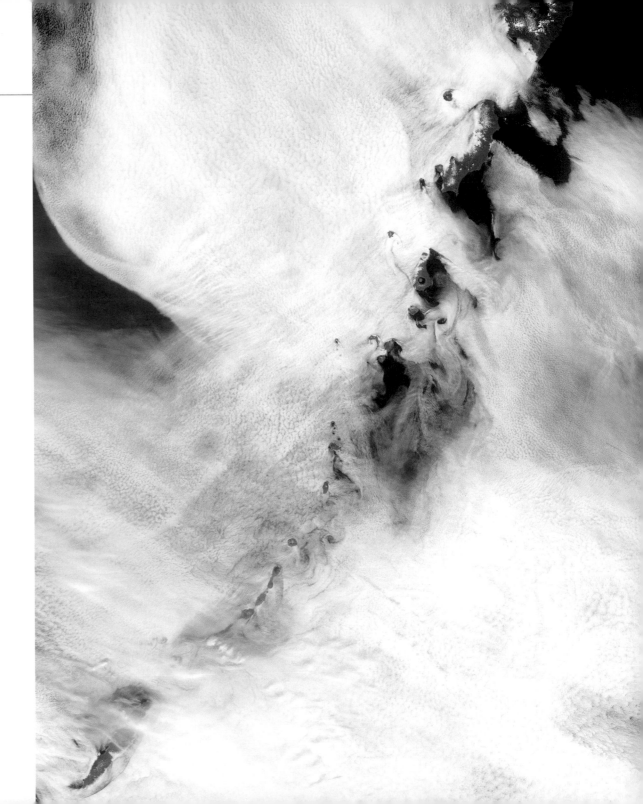

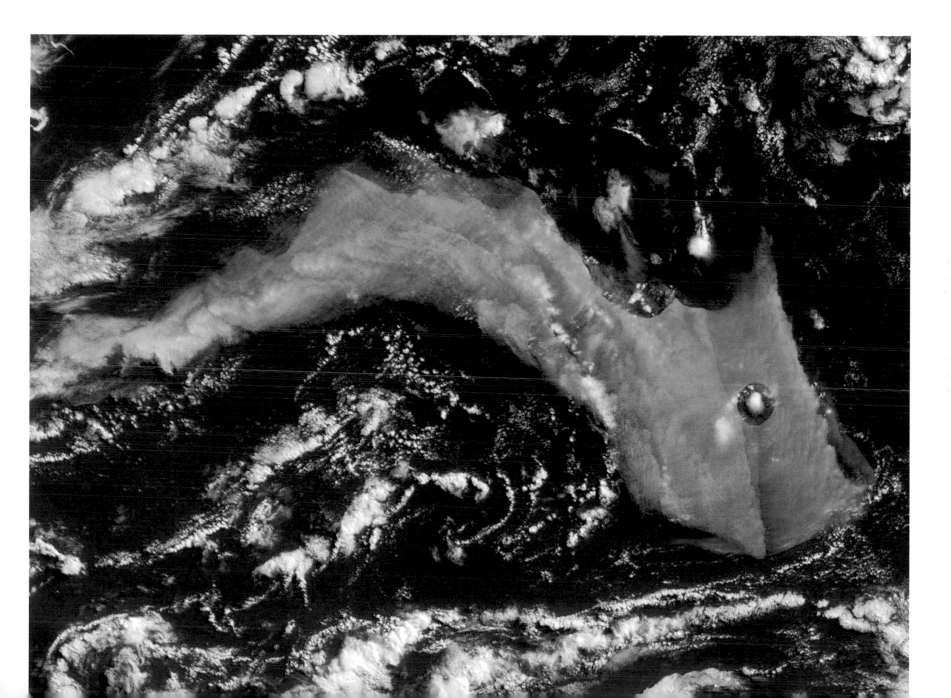

"IT IS A PERMANENT CHALLENGE
TO ANTICIPATE WHEN OR WHERE
TROPICAL CYCLONES WILL
THREATEN PUBLIC SAFETY"

CHAPTER THREE:
HURRICANES AND DRAMATIC WEATHER

When we talk about storms, we usually mean thunderstorms. These are often quite small storms borne in individual clouds or small clusters. But, over warm tropical waters, storms are born on an altogether grander scale, and might be embedded with many thunderstorm cells.

Hot, moisture-laden air rises from the sea to create huge areas of deep low pressure – tropical cyclones. In the Atlantic Ocean, tropical cyclones are called hurricanes, but in the Pacific are generally known as typhoons. In the Indian Ocean, they are called severe cyclonic storms or tropical cyclones. The key features that they all share are massive rotating cloud structures, intense rainfall and very strong winds. Curiously, there is often a quiet, cloudless eye in the centre of the storm.

Once tropical cyclones have developed they often intensify dramatically in just a few days, but when they make landfall rapidly weaken. It is difficult to forecast their precise speed and track, and it remains a permanent challenge to anticipate when or where they will threaten public safety. Consequently, research continues into improving forecasting techniques for tropical cyclones. Hurricanes, cyclones, and typhoons are all seen in this chapter.

TROPICAL CYCLONES

◄ CYCLONE CYPRIEN

Cyclone Cyprien, shown here on 1 January, 2002, brought massive rainfall totals of 10–18cm (4–7in) over southern parts of Madagascar. The cyclone was heading slowly towards the eastern coast of Africa. The winds blow clockwise around low pressure, creating a shape unfamiliar to those living in the northern hemisphere.

TROPICAL CYCLONE DINA ➤

On 21 January, 2002, this image catches Tropical Cyclone Dina northeast of Mauritius and Reunion Island. The storm's centre is marked by the eye, where winds are usually light and the sky often clear of cloud. At this stage, Dina had already generated winds of 220km/h (138mph) with gusts exceeding 260km/h (160mph) and was still developing.

SUPER TYPHOON PODUL ➤➤

A Super Typhoon is one with sustained winds of at least 240km/h (150mph). This image of Super Typhoon Podul, just east of the Mariana Islands in the western Pacific, captures it when wind speeds were reaching a massive 270km/h (167mph) with gusts to 306km/h (190mph). It was one of the most severe storms of that season.

CYCLONES/
DEPRESSIONS

◄ MULTIPLE CYCLONES IN THE INDIAN OCEAN

This train of four tropical cyclones in the southern Indian Ocean was captured on camera on 11–12 February, 2003. The cyclones are called (from the left) Gerry, Hape, 18S, and Fiona. This picture is what is called a composite image, and is made up from four images taken from alternating passes of Aqua and Terra, two orbiting satellites. Orbiting satellites go round the Earth several times a day; each picture taken over a particular area is known as a pass.

SWIRLING MASK OF LOW PRESSURE TWINS ➤

The two swirling cloud masses in this picture are depressions (areas of low pressure). This pair was captured in November 2006 southeast and southwest of Iceland in the North Atlantic; the island is visible as a ragged white area at top centre, with Scotland in the extreme bottom right-hand corner. The rope-like streams are frontal cloud systems, swirling into the centre of each depression, the cloud from the stronger feature (right) lying over that from the smaller low (left). To the south of them both lies a wonderful cloudscape of open convective shower clouds, some in streets.

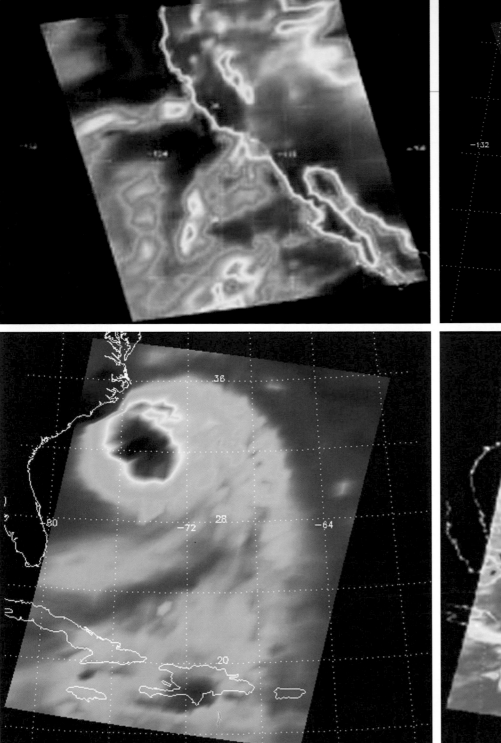

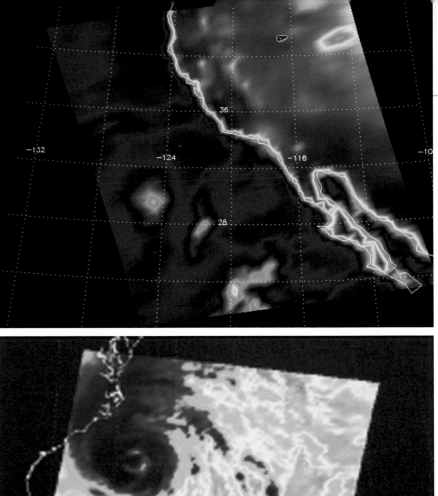

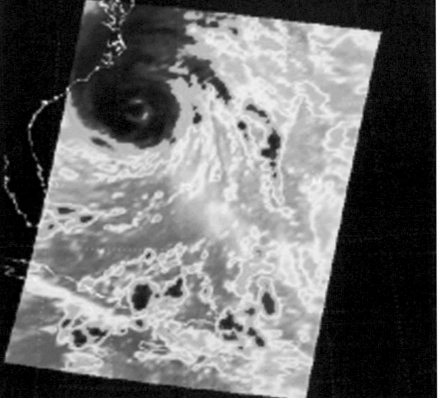

STORM
MONITORING

◄ DIFFERENT SATELLITE IMAGING SENSORS

With visible, infrared, and microwave detectors, satellites can provide an amazing 3D look at the Earth's weather, even with heavy clouds. The upper set of images is of a storm approaching California. The one on the left uses colours to grade temperatures (purple is cold, red is warm) while the one on the right measures water-vapour content and land-surface temperature. We can see some intense pockets of moisture and a hot land mass. The lower pair shows Hurricane Isabel just off the southern coast of the USA. The left-hand image shows the water vapour in the atmosphere ranging from the intense red (very moist) at the hurricane's eye to the pale blue regions of less humid air further out, while the right-hand image shows the temperature of the cloud tops, blue being very cold (high).

The use of multiple images from different ➤ satellite sensors gives forecasters vital insight into how a weather system is developing. This image shows a prominent squall line, pointing nearly north–south, approaching the California coast with a large cloud formation almost due west. Both features have high cold cloud tops, according to the blue/purple images, and both were probably a major source of intense rainfall.

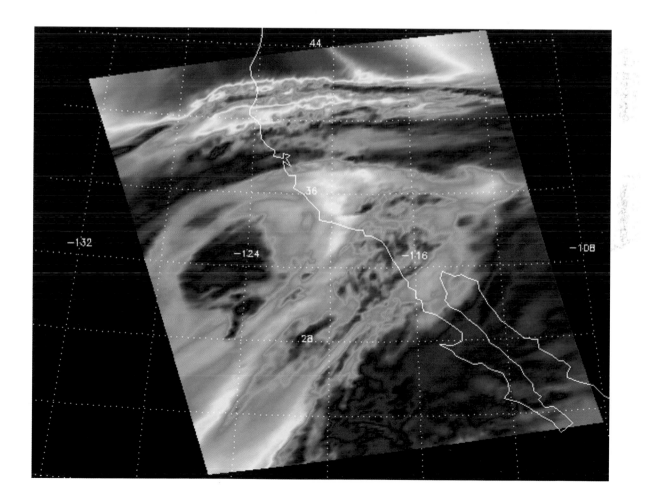

26° forward

0   10   20

height in kilometers

◄ DOUBLE CLOUD IMAGES OF HURRICANE JULIETTE

On the left we have a single-instrument, true-colour image of the intense Hurricane Juliette taken on 26 September, 2001. The spiral arms around the hurricane's clearly visible eye are impressive and show several areas of lumpy, embedded thunderstorm clouds within them. On the right, the coloured image of the cloud-top height is computer-derived from several cameras. The highest clouds, and the thickest, show as bright orange while lower, thinner clouds show as green or even blue. There is a patch of blue/green within the hurricane's eye.

MONITORING HURRICANE WILMA, OCTOBER 2005 ➤

Hurricane Wilma was the 21st named storm of the 2005 Atlantic hurricane season. It developed very quickly, striking Mexico's Yucatán Peninsula as a storm of category four before finally reaching Florida at category three. Each of these three pairs of images shows a traditional picture alongside one displaying cloud-top height in false colour, from the left, the hurricane developing, hitting Yucatán, and reaching Florida. Red and orange indicate very high, thick cloud; purples are low. Note the boiling, lumpy cloud in the left grey image with corresponding bright red/orange in the cloud height, and the distinct eye in the central pair. On the right, Wilma's eastern flank can be seen finally weakening.

Tuesday, October 18   0   9   18   height in kilometers

Friday, October 21   0   9   18   height in kilometers

Sunday, October 23   0   9   18   height in kilometers

STORM SYSTEMS

◄ HURRICANE CLAUDETTE FROM THE INTERNATIONAL SPACE STATION

These two photographs provide a wonderfully contrasting pair of images of the same event. Initially a tropical storm, Claudette developed to become a category one hurricane, the first of the 2003 Atlantic season. This photograph (left) was taken by an astronaut on the International Space Station on 15 July just as Claudette made landfall and started to weaken. It still managed winds of 130km/h (80mph) and gave that part of Texas, including the astronauts' home base in Houston, a real soaking.

TROPICAL STORM CLAUDETTE OVER TEXAS AND MEXICO ►

By the time of this shot (right) the following day, the cloud is thinning and breaking up, and the storm is no longer classified as a hurricane. It still shows the characteristic spiral cloud pattern of a tropical storm, however, and is centred over the west Texas Panhandle-Mexico border. This image also shows some tiny spots of red in the top left-hand corner. These are fires in New Mexico and Arizona, with smoke drifting away from them.

◄◄ TROPICAL CYCLONE FAY OVER WESTERN AUSTRALIA

This image of tropical cyclone Fay skating down the western Australian coast on 26 March, 2004, creates a dramatic contrast between the white cloud mass, the brown land, and hints of the shallow water just offshore. With winds of 166km/h (103 mph), the Australian Bureau of Meteorology classified Fay as a severe tropical cyclone; they ranked it a category four storm, category five being the most intense.

◄ FAMILY OF THREE PACIFIC TYPHOONS

This photograph captures an unusual, but by no means unprecedented, occurrence of three storm systems in the same general area at the same time. They are Bopha (left), Maria (top right), and Saomai (bottom right). Taken around midday local time on 7 August 2006, Bopha is only a few hours old with just the basic round shape of a tropical depression. Maria, a day older, shows a more distinct spiral structure with extending arms of cloud, while the even older and much more powerful Saomai shows a closed eye.

GLOBAL VIEWS
OF HURRICANES

◄ HURRICANE FRANCES 2004

This image of Hurricane Frances on 31 August, 2004, shows the storm about one-third of the way through its life. It was a fearsome hurricane, with sustained winds up to 235km/h (145mph). The eye of the hurricane passed over San Salvador Island and was very close to the Bahamas, while Frances' outer bands affected Puerto Rico and the British Virgin Islands. The hurricane made landfall in central Florida, crossed the state, curved round the Gulf of Mexico, and finally made a second landfall at the Florida Panhandle. Total damage was estimated at around $10 billion, with seven direct fatalities.

HURRICANE LINDA 1997 ►

This image is truly unique, a one-off combination of science, engineering, and artistry, combining digital weather, ocean, and land colour data from three satellites over a period in September/October 1997. It shows Hurricane Linda just off the west coast of North America and highlights such features as the Amazon River, including the sediments around the river mouth.

THE TOP OF THE ATMOSPHERE

This astronaut's view on 20 July 2006, captures
the otherwise invisible boundary between the
Earth's atmosphere and outer space. Above
the cloud cover is a blue halo caused by the
scattering of blue wavelengths of visible light
more than others, the same effect that makes
our sky appear blue. And an almost translucent
moon floats between Earth and space. Although
there is no real boundary, we think of the top
of the atmosphere as being about 100km (60
miles) above the Earth. This marks the level
at which solar energy enters the atmosphere,
and where the balance between incoming and
outgoing energy affects the Earth's average
temperature.

◄ AEROSOLS OVER INDIA

In this image, taken on 4 December 2001, a thick soup of aerosol particles along the southern edge of the Himalayan Mountains, and streaming southward over Bangladesh and the Bay of Bengal, contrasts starkly with the pristine air over the Tibetan Plateau to the north of the Himalayas. This rich chemical mix of pollutants has an impact on both the health of the local people and on the region's rainfall cycle and climate. The Brahmaputra (upper right) and Ganges Rivers are still visible, the latter's many mouths having emptied their sediment-laden waters into the bay and turned the northern waters of the Bay of Bengal a murky brown.

AEROSOLS OVER THE YELLOW RIVER VALLEY ➤➤

Thick aerosols blanket much of China's Yellow River Valley in this image from 22 October 2001. Running briefly east from the upper left, the valley extends south before turning east again near the bottom centre of the image and before the cloud flows out across the plain into Bo Hai Bay. In the centre of the image, aerosols are packed into the Fen River Valley, produced by its many cities. What kept the aerosols confined to the lower levels on this day was a strong temperature inversion that trapped the cool dense polluted air beneath an invisible lid.

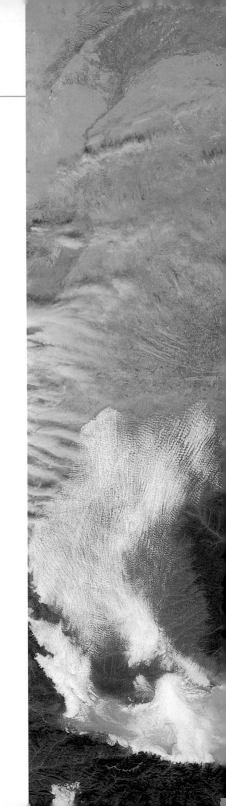

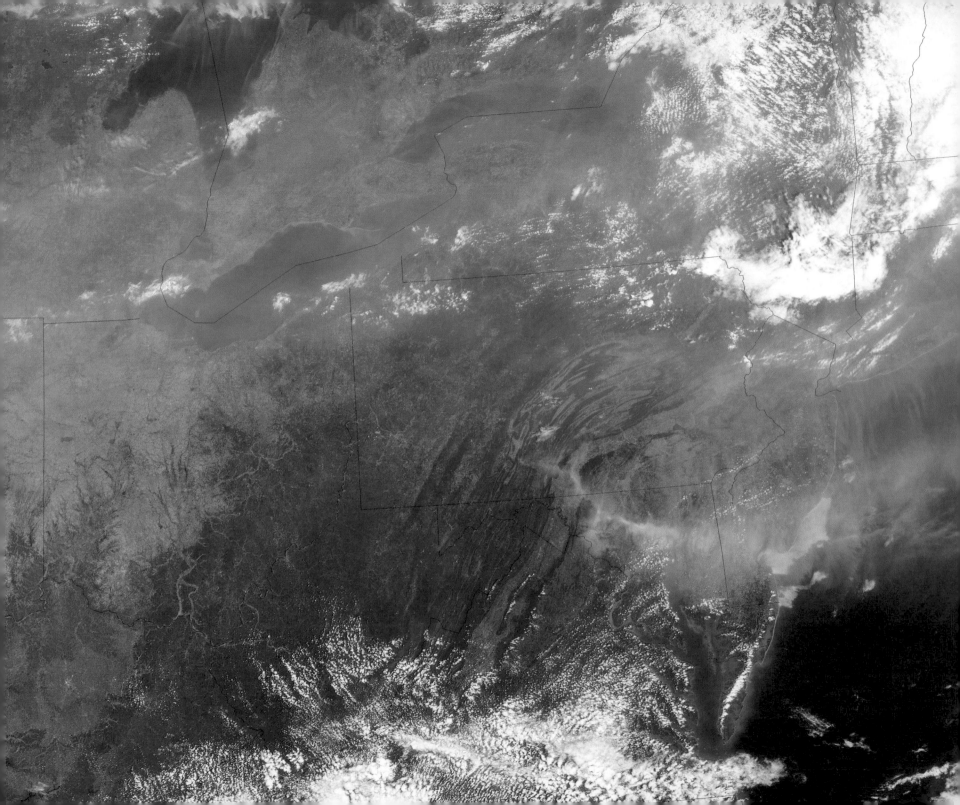

◄◄ POLLUTION OVER THE GREAT LAKES AND US EAST COAST

Air quality across the Great Lakes and mid-Atlantic regions of the United States is less than ideal in late June 2002. Much of the pollution is smoke from forest fires in the west of the USA and fires in Canada's Prairie Provinces to the north. Since the dominant weather patterns across the country move air from west to east, it is not uncommon for air quality along the East Coast to be affected by fires out West. What's more, you may be able to spot some tiny red dots in the centre of the image. These are fires in the eastern states, and they provide even more haze.

◄ AEROSOLS OVER MID-ATLANTIC UNITED STATES

The extensive haze in this true-colour MODIS image is probably a mixture of smoke from fires burning in Virginia and North Carolina and air pollution in the mid-Atlantic United States. Such fires are not uncommon over North America during the later summer months, and atmospheric pollution of varying degrees is simply a fact of life for the industrial USA. Both contribute to the increase in greenhouse gases which lead to climate change globally.

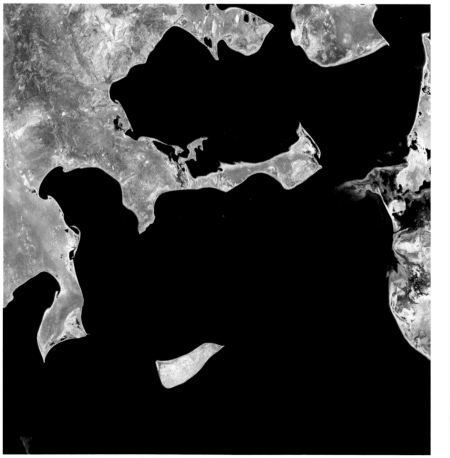
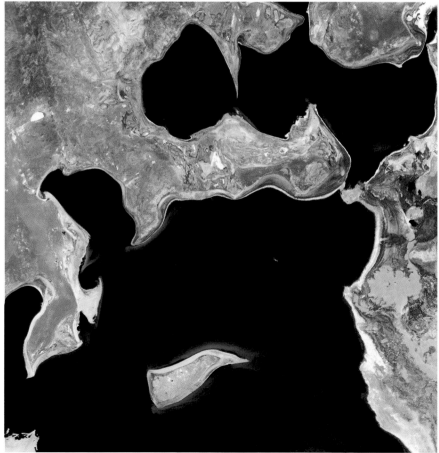

THE SHRINKING ARAL SEA

The Aral Sea is actually not a sea at all. It is (or was) the world's biggest lake, a body of fresh water. Its shrinkage is not due to climate change but to another aspect of mankind's impact on the environment: the excessive stealing of water to irrigate cotton fields and rice paddies.

From left to right the pictures show the Aral Sea in 1973, 1987 and 2000. During this time more than 60 per cent of the lake's water, shown in black, disappeared. With a massive decrease of in-flowing water, concentrations of salts and minerals have soared, causing precipitous drops in the fish population and the demise of commercial fishing. Strong winds strip away tons of exposed soil, causing poor air quality and reducing crop yields, an unsustainable yet ongoing position.

◄ SAHARAN DUST STORM THREATENS LAKE CHAD

In the south-central reaches of northern Africa's Sahara Desert, a storm is whipping up sand and shifting it southwards over an already-struggling Lake Chad (left centre edge). A mixture of increased water demands and lower rainfall, possibly from climate change, have reduced the lake's surface area dramatically. Now a ghost of its former self, it has shrunk to one-twentieth of the size it was 35 years ago.

THE OPTIMIST, KALAHARI DESERT, NAMIBIA ►

This is another image linked to encroaching desert sand. Looking rather like a piece of modern art, the sensor onboard satellite Landsat 7 has produced this banded image of vegetation and land use. Healthy vegetation appears red. Sadly, there is little red left as sand dunes spread onto once fertile land. But there is a lone bright red spot (upper centre) denoting at least one optimistic farmer, using an irrigation system, who continues to farm despite the approaching sand.

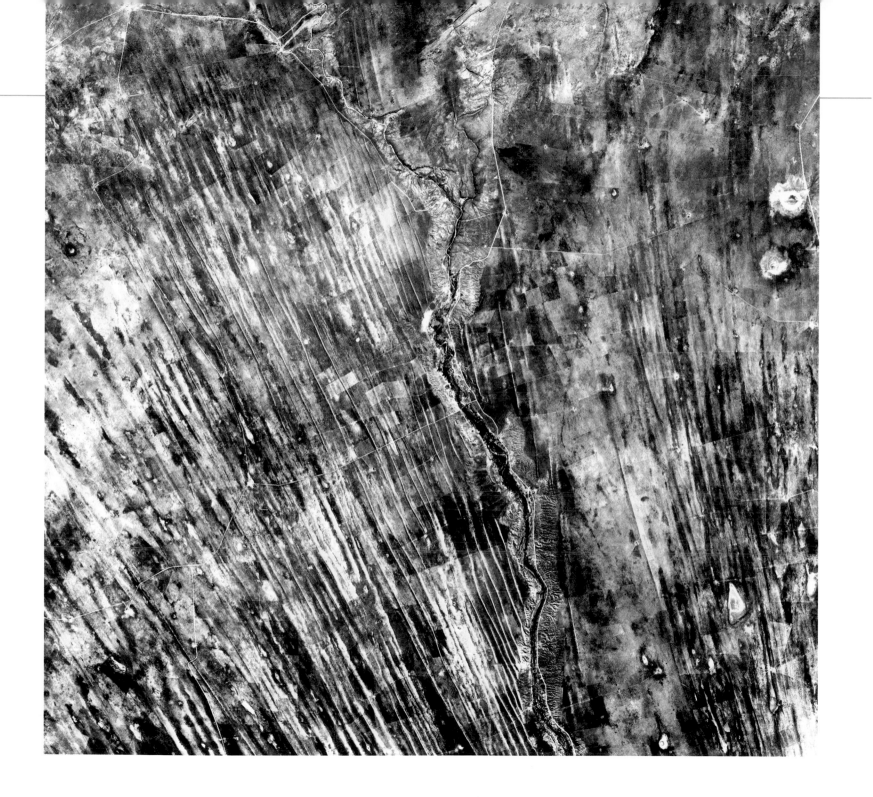

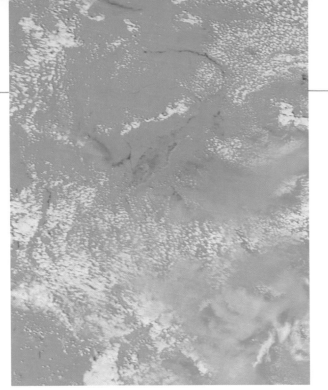

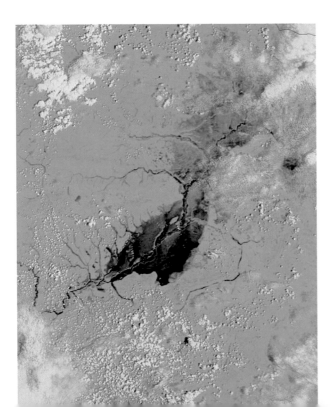

◀ CHAMBESHI RIVER FLOODS, ZAMBIA

Many rivers in southern Africa, including the Zambezi, broke their banks in early 2007 following an exceptionally early and heavy rainy season. This pair of images from January and February shows the Chambeshi River in Zambia, a source of the mighty Congo River, before and after the rains. In these pictures, water is shown as black, vegetation as green, bare soil as tan, and cloud as white or light blue. What was barely a visible line in the upper picture has become a broad lake over 20km (12 miles) across in the lower one. In east Zambia alone more than 23,500 households were affected by flooding.

FLOODING IN BANGLADESH ➤

North-eastern Bangladesh disappeared under monsoon floods as rain drenched the region in June 2006. By the date of this image on 21 June, hundreds of square kilometres of what had been dry land just a month earlier were under water, shown black or dark blue. Plant-covered land shows green and clouds are white or pale blue. The floods claimed nine lives, made 10,000 people homeless, and left more than a million stranded. With sea levels expected to rise, the disappearance of habitable land in southern Bangladesh will become an increasing problem, too.

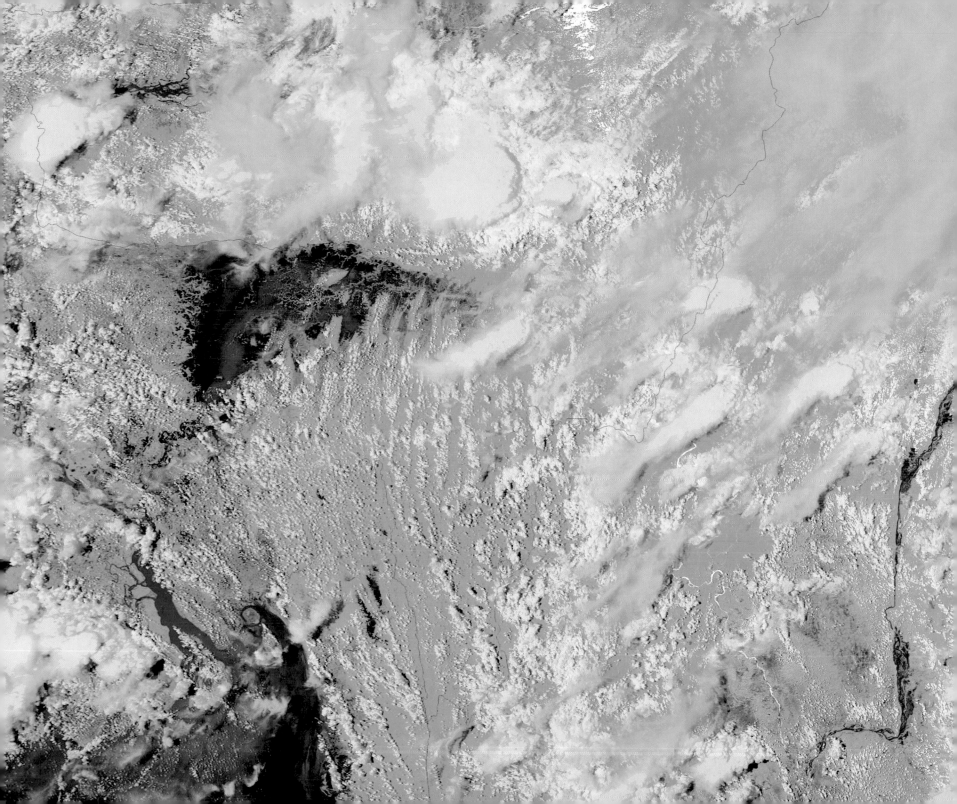

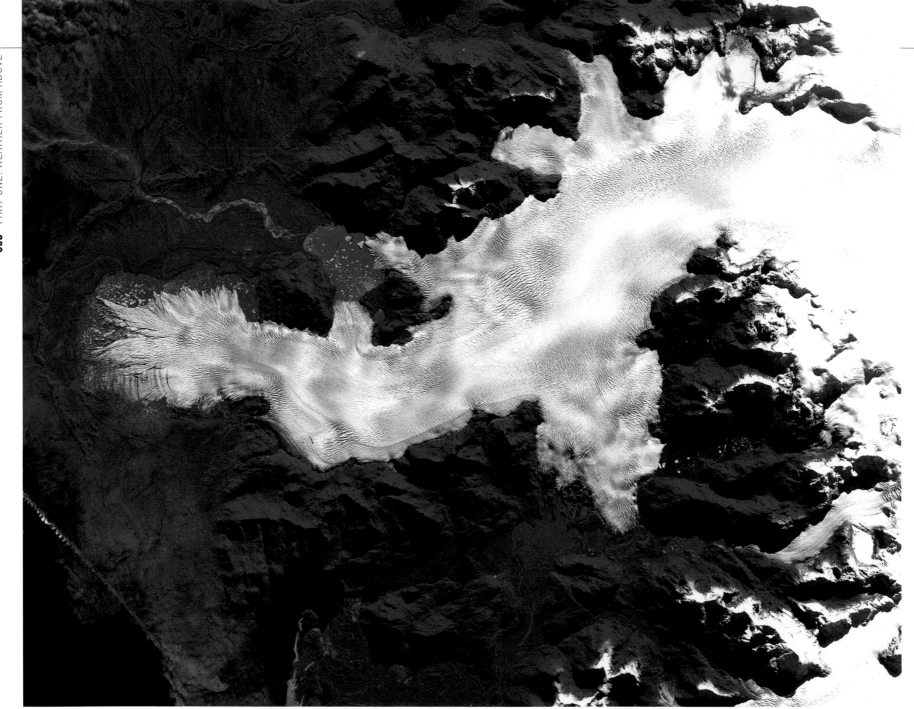

◄◄ PATAGONIA GLACIER, CHILE

The decrease in size, or retreat, of glaciers is well recognized as an indicator of global warming. Patagonia is a mountainous region spanning the border between Chile and Argentina near the southern tip of South America. In this image of a Patagonian glacier, vegetation is coloured red and the ice is bright white. By monitoring the size of glaciers such as this one, scientists can gather categorical evidence regarding global warming.

◄ PASTERZE GLACIER IN AUSTRIA

This west Austrian alpine valley glacier, the Pasterze, has been retreating since 1856. This has been caused by a combination of lower winter snowfall and higher summer temperatures. Land that was covered in glacial ice just 10 years ago is now just rocks and water run-off. In 2003, glaciers in next-door Switzerland receded more quickly than in any other year since their records began in 1880. The Swiss Academy of Natural Sciences attributes the melting and recession to long-term climate change.

◄ DETAILED VIEW OF SEA ICE

Snow and ice are not just the result of a cold climate; they also help create it. So if the climate warms up and melts them, their loss plays a part in speeding this process along. This influential effect can be seen in images like this one of Arctic sea ice captured on 16 June, 2001, by NASA's Landsat 7 satellite. Sea ice is very white and reflects incoming solar energy; the Arctic waters are very dark and absorb energy. A warming climate causes ice and snow to melt, which increases the sea surface and solar absorption, leading to more warming.

HELHEIM GLACIER RETREAT ►

This pair of images shows the retreat of the Helheim glacier as it flows off the Greenland Ice Sheet. From the 1970s until about 2001, the glacier's front edge moved very little, but between 2001 and 2005, it retreated almost 7.5 km (4.5 miles), easily seen if you compare the position of the glacier's edge to the two vertical valley formations (top 2003, bottom 2005). Moreover, it thinned by about 40m (130ft) in the two years 2001 to 2003. Part of the thinning is because of warmer temperatures, but changes in glacier speed also play their part. And this pattern of melting, thinning and recession is seen all over the Greenland Ice Sheet.

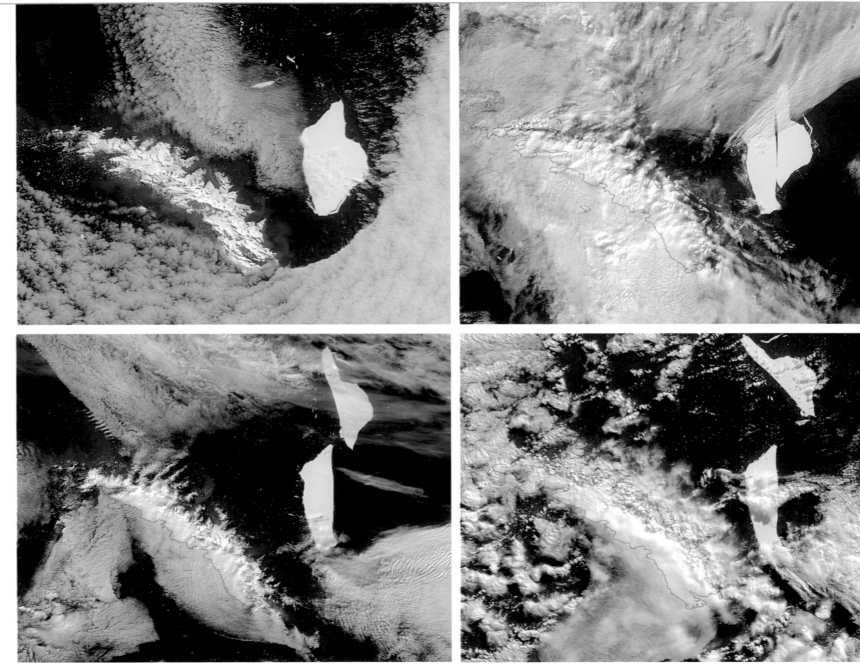

◄◄ GIANT ICEBERG SPLITS

On 13 October 1998, the biggest iceberg
seen for a decade broke off the Ronne Ice
Shelf, Antarctica; scientists labelled it A-38.
By 22 October, a big piece broke away and this
was labelled A-38B; it drifted some 2400km
(1500 miles) north to near South Georgia in
the South Atlantic. On 12 April 2004, (top
left), A-38B was 40km (25 miles) long; by 15
April (top right), it had split in half. By 17 and
18 April (bottom left and right) one part had
moved quickly north and turned while the other
seemed fixed in place. Another ice-shelf berg
was on its way to melting completely

◄ THE MELTING GREENLAND ICE SHEET

Every year a little more of the Greenland Ice
Sheet melts away, but September 2002 was
a record melting month. There is a seasonal
melting every summer but it is the sheer scale
of the melting that is now leading to a reduction
in the size and depth of the ice. The melting
process is quite complex but as melting water
collects, it seeps down through the ice until
it reaches the land surface beneath. Here
it loosens the bond between land and ice,
speeding up the breakdown process. In this
image, unusually large areas of brown land are
visible.

PART TWO: WEATHER FROM BELOW

We move now from space down to the Earth. Most of the images in this second section of *Weather World* are taken from the Earth looking up at the sights in the sky. Some are easily spotted, some very rare. Some are hazy and ephemeral, some are spectacularly bold and bright. From clouds to rainbows to lightning, here are some of the weather's most wonderful and bizarre manifestations.

CUMULONIMBUS

CLASSIC ANVIL ➤
An anvil forms when the stronger winds higher in the atmosphere catch hold of the cloud top and sweep it forwards. This might not look like much of a cloud, but when you consider it formed over the soaring mountain range in the distance, you have to think again. Captured in French, New Mexico, it has a high base, which usually means the low-level air is relatively dry, while the top of the anvil is still predominantly made up of water droplets. Because the top of the cloud is shearing forwards, it creates a spectacular anvil in the sky.

◄ CUMULONIMBUS IN KANSAS
This cumulonimbus cloud picture was shot in Goodland, Kansas. There is not a lot of power driving the cloud; it appears to be late in the day and the Sun's heat is fading, restricting the cloud's growth. This one is not likely to become a giant thunderstorm.

◄ RED SAIL IN THE SUNSET

This luminous evening image of a
cumulonimbus is taken from Beadnell,
Northumberland, in northern England.
The cloud is a long way off, and has a relatively
low top. This is typical of a winter, rather than a
summer, thunderstorm. It has probably formed
in a cold, unstable northerly air stream, the
warmer sea acting as the source of heat that
has triggered the cloud.

BORNEO BEAUTY ➤

This shot was captured in Sabah, Borneo. The
striking hues reflect the time of day, towards
sunset, when the evening sky darkens the cloud
away from the sun's light and creates vivid
yellows or golds where the sun's last rays still
light up the cloud top. As the sun starts to set,
the vigorous clouds lose the power source that
drives them and they begin to decline. This one
is just at the point of dying away and will soon
spread out into other cloud shapes.

◄ A STORM IS BORN

At the point of creating an anvil, you can almost feel the currents of air boiling upwards and driving this cumulonimbus in Goodland, Kansas. The air currents have taken the cloud to the height where water droplets give way to ice crystals, and so the top of the cloud is starting to glaciate. You can clearly see the different textures of the lower cauliflower heads of the larger cumulus clouds and the fibrous look of the main cloud top. This cloud will probably go on to produce a fully formed anvil and then a thunderstorm.

TEXAS BOILER ►

This boiling pot of action was captured in Throckmorton, Texas. The white cumulonimbus in the middle of the picture is the main focus, and was almost certainly born out of the vigorous updraughts, indicated by the dark cloud masses around it. This was a stormy day with the atmosphere in real turmoil.

◄ TWIN PEAKS

The dual top to this cloud suggests a significant heat source on the ground, which has powered it way beyond the general cloud-top level. It may have been a particular southward-facing hillside or an area of rock. Either way it has created a powerful, narrow column of air, and then produced these two cloud heads. It looks as if the very top of the one on the left has drifted downwind of the rising air, been cut off from the updraught, and the second head has then taken its place. This one was shot in Harper County, Texas.

MULTI-CELLED CUMULONIMBUS ►

There are several cumulonimbus cloud cells powering up in this picture from Dimmitt, Texas. The one in the foreground is a new cell, born out of downdraughts of air from the main cloud behind it. There is also the effect of the surging updraughts reaching the tropopause – the top of the troposphere, the layer of the atmosphere in which most of our weather occurs – and spreading out, front and back, before cooling and allowing the air to fall back towards the ground. It is these downdraughts of air that often trigger the formation of another cloud cell or even a tornado.

◄◄ THE ROAD TO TROUBLE

This black-top South Dakota road splitting the agricultural landscape is heading right into heavy rain at the heart of a massive cumulonimbus cloud. On the underside of the main cumulonimbus cloud are mamma or mammatus clouds of varying sizes, so called because of their breast-like shapes. Their presence indicates very strong up- and downdraughts within the thunderstorm cloud.

◄ OKLAHOMA SUPER-CELL

This remarkable image of a cumulonimbus cloud over Woodward, Oklahoma, is partly due to the evening sunlight and partly to the massive depth of the cloud. It is almost certainly a super-cell cloud, one that contains a consistent, single, strong, rotating updraught. These clouds produce high winds and large hailstones, and lead to the formation of tornadoes. Although no anvil is visible, it is clearly a cloud of great height, probably soaring to some 12,000m (40,000ft) or so. The lower part of the cloud (towards the right) is a wall cloud, often a feature of super-cell storms.

CUMULUS

TOWERING CUMULUS ➤

These large cumulus clouds are known as
cumulus congestus, the very phrase suggesting
something solid and full. They were captured
from an aircraft flying over Ecuador, and clearly
have a bubbling, building feel about them,
particularly the largest cloud. This towering
cumulus can grow into something larger,
perhaps even a cumulonimbus.

CARIBBEAN CONVECTION ➤➤

If you have ever looked up into the sky on a
day when convection is really vigorous, when
the air is unstable and ready to rise with a
minimum of heating or lifting, then you may
have seen clouds like these. Positively boiling
with energy, you can watch them actually
growing and changing shape. Smaller cousins of
cumulonimbus, these cumulus congestus were
captured at Morgan Lewis Bay on the Caribbean
island of Barbados.

CUMULUS CONGESTUS

These cumulus congestus clouds, captured
in French, New Mexico, are almost certainly
on their way to turning into cumulonimbus.
There are the earliest signs of ice crystals in
the very tops of the clouds, a sure sign that a
cumulonimbus is starting to form. The cloud
head at the top right is taking on a pre-anvil
shape, while beneath the base of the main
cloud there are the characteristic dark hues
and shades of a rain shower.

◄ GOOD EVENING CUMULUS

This evening display of large cumulus clouds
was seen in the county of Northumberland,
in northern England. An intensely blue sky,
vigorous clouds, and amazing light effects
produce a very vivid image. The clouds are
still vigorous despite the lateness of the day,
suggesting a really unstable air stream.

CUMULUS IN SILHOUETTE ►

Evening hues show these towering cumulus
clouds off to spectacular effect. Captured in
Borneo, this image shows the clouds late in the
day, but still with enough energy in the rising
air columns to sustain their bulging tops. As the
sun finally sets, local heating will die away, and
the clouds will slowly spread out and disperse.

EIGHT INTO ONE ➤

The multiple towers of Thrumpton Power
Station in Nottinghamshire, England, combine
to produce this single pyrocumulus cloud. The
day is still, the rising air is undisturbed, and
the cloud itself is fairly flimsy. This suggests
the surrounding air is dry and not too much
moisture is coming from the cooling towers
themselves. These man-made cumulus clouds
are more common than you might think and,
when they look like this one, neatly illustrate the
way a cloud is formed.

STREETS IN THE LANDSCAPE ➤➤

These cumulus cloud streets have formed
between the lake to the south of East Grinstead,
West Sussex, England, and the town itself. The
main street is like a progression of pale stepping
stones right over the lake, and with a particularly
large clear area of sky to the south. The air flow
has created two organized streets and there are
plenty of other scattered small cumulus clouds
over the lower hills.

STRATOCUMULUS

WINTER STRATOCUMULUS

This is a typical sheet of moist stratocumulus,
a cloud that forms when the atmosphere
is relatively stable with an inversion, when
temperature rises with height instead of falling,
acting as a limit to the vertical development
of clouds. It usually displays a slightly bumpy
base, as in this case, with individual elements
all merging together. But it can also occur in

◂◂ EVENING STRATOCUMULUS

In this picture, we can see the many individual elements of a stratocumulus layer, and realize it is not a solid layer at all. The elements have a puffy look, more like cumulus, which, in a sense, they are. Each element is like a very small cumulus cloud that cannot grow because of the temperature inversion above it; this acts like a lid and causes the clouds to spread out.

◂ STRATOCUMULUS CASTELLANUS

These fluffy cotton-wool balls of stratocumulus, captured in Edgefield, Norfolk, England are stratocumulus castellanus. They could be mistaken for fully fledged small cumulus clouds, but they are not created by pure convection. There is a lot of turbulent mixing involved in the creation of stratocumulus, and there is enough instability in the atmosphere to allow these elements to puff up into the individual lumpy shapes we see here.

◄ ALMOST GONE...

This lovely sunset really highlights the characteristic rippled base and shallow nature of stratocumulus. While this shot shows stratocumulus layers, these clouds can also form from the spreading out of cumulus clouds late in the day when convection dies away. The long strip close to the horizon is just such a cloud.

THE CLOSE OF THE DAY ➤

Another lovely sunset – the end of a peaceful day, weather-wise. Captured in southern England, we can see the small cloud elements in this stratocumulus layer, many of them still catching the pure white light of the sun. Lower down on the horizon, the underside of the clouds is catching the golden orange and red hues of the setting sun.

STRATUS

LOST JUNGLE

Fog-like stratus clouds have created the mysterious look of this jungle on the lower slopes of Mount Batur in Bali, Indonesia. This is tropical or equatorial rain forest, where it is very warm and humid, and the clouds are pierced by the jungle's giant trees. But this effect can also be seen in cold damp parts of the world too, as we will see.

◄ MURKY MOUNTAINS

In contrast to the picture on the previous page, here we see two examples of stratus fractus, broken low cloud, in a cold setting. Cloud formation happens when the temperature cools to the level at which water vapour in the air condenses into water droplets. The shot (top, left) was taken on Ben Nevis, Scotland, and shows the low cloud, familiar to hill walkers and mountaineers, close to the lochan part of the way up the UK's highest mountain. And the picture (bottom, left) is of stratus drifting over another Scottish area with a murky historical past, Glen Coe, in Argyll.

STRATUS IN THE SNOW ►

On the right, stratus has formed over the cold alpine resort of Chamonix, in France. The featureless cloud has crept down the hillside, providing a misty ceiling. The cable car will no doubt disappear into the cloud quite soon. Not good weather for skiing.

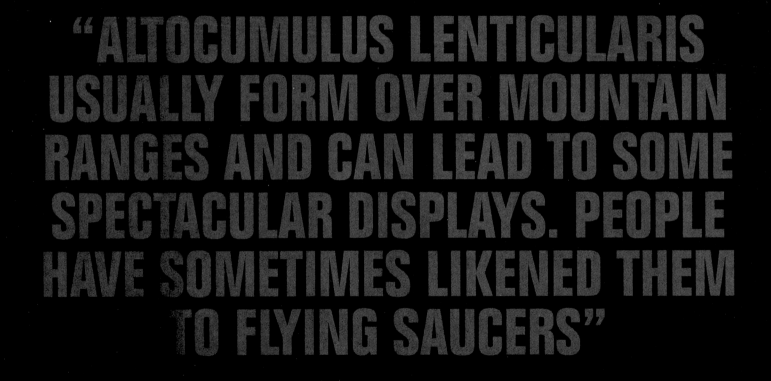

"ALTOCUMULUS LENTICULARIS USUALLY FORM OVER MOUNTAIN RANGES AND CAN LEAD TO SOME SPECTACULAR DISPLAYS. PEOPLE HAVE SOMETIMES LIKENED THEM TO FLYING SAUCERS"

CHAPTER SIX:
MEDIUM CLOUDS

Medium clouds are those that form between 2,000m (6,500ft) and 6,000m (20,000ft). They are largely made of water droplets, which produce clouds of two main varieties: the sheet-like altostratus, and the more attractive fleecy altocumulus, which creates the characteristic mackerel sky look. There is also one cloud type, nimbostratus, that is really a type of low cloud but is thought of as medium cloud. It is typically thick, grey, and featureless, and can produce heavy rain.

Altostratus is not very pretty; it is simply a bland layer of grey. What can make it more attractive is the effect of the Sun or Moon shining through the cloud, illuminating an area of the sky. Depending on the size of the cloud water droplets, the illuminated area can be coloured, producing what is known as a corona (see Chapter Nine).

Altocumulus, on the other hand, takes many beautiful shapes and patterns, varying from the familiar sheep fleeces to the more unusual lentil-shaped clouds, known as lenticularis. These usually form over mountain ranges and can lead to some spectacular displays. People have sometimes likened them to flying saucers.

NIMBOSTRATUS

◄ A THICK WET LAYER

Nimbostratus is a bad-weather cloud, nearly always bringing rain or snow. It usually occurs as a dark grey, featureless layer, but can have a somewhat ragged base. Indeed, it is often accompanied by patches of even lower stratus fractus clouds. One feature of the cloud by day is its eerie internal lighting. It emits a kind of glow as the Sun does its best to pierce the thick cloud.

◄ BAD WEATHER TO COME

There is just a hint of brightness penetrating this combination of nimbostratus with stratus patches beneath. These grey amorphous layered clouds are among the most difficult to catch photographically, and never look very exciting. That said, the wet or snowy weather they bring promises very interesting conditions: prolonged days of rain, flooded rivers, or massive snow falls. Dull and grey they may be, but they are a very real warning of dramatic weather change anywhere in the world.

ALTOSTRATUS

◄ ALTOSTRATUS AT ITS BEST

This gorgeous sunset in Berkshire, England, belies the usual grey, layered nature of this altostratus cloud. Altostratus takes two forms: thin and thick. Altostratus translucidus lets more light through than the thicker altostratus opacus. This is an example of altostratus translucidus at its very best, with the whole sheet of cloud lighting up in the western sky.

◄◄ VARIED ALTOSTRATUS OPACUS

Altostratus is a layer cloud typical of an
approaching warm front. It usually thickens
with time, turning from thin to thick, and then
often into nimbostratus. In any event, it usually
means rain is not far away. This is a layer of thick
altostratus over southern England. It has a base
around 2,500m (8,000ft), leaving plenty of room
beneath it for ragged patches of stratus pannus;
these are probably around 450m (1,500ft).

The second shot on the left (below) is another
example of thick altostratus, though this time it
appears in the sky on its own with a rather more
varied texture and colour.

◄ A FALL-STREAK HOLE IN THE SKY

Medium clouds are composed mainly of
super-cooled water droplets, those that are
well below freezing point but have not actually
frozen. In this image, caught at Totland on the
Isle of Wight, UK, an unknown phenomenon
has caused the super-cooled droplets to freeze
and then fall, disappearing as they do, a process
called sublimation. When this happens we are
left looking at a fall-streak hole in the main cloud
layer and vestigial ice-crystal cloud.

◄◄ ISLE OF WIGHT FALL-STREAK HOLE

This image has captured a thick layer of altocumulus with a fall-streak hole in it. Very thin altocumulus cloud elements can be seen in the hole, with a pretty patch of cirrus cloud well above. The striking contrast between the bright white clouds and the deep blue of the sky creates an effect just like looking through a hole into deep blue water.

◄ EERIE FACES IN THE SKY

These two remarkable cloud structures are a form of altocumulus lenticularis, the lens-shaped clouds created by wind flow over mountain barriers. Some aspect of this air-flow has produced dramatic horizontal swirls on the undersides of the usually smooth clouds and created eerie faces, which could be mistaken for flying saucers.

◄ ANTARCTIC ALTOCUMULUS

Cloud formations are very similar all over the world. This example of a thin sheet of altocumulus stratiformis was caught at Rothera in Antarctica, one of the most important British research bases on the continent. The image is a study in shades of blue, grey, and white, as the cloud is reflected in the icy waters alongside the icebergs. The sharp edge of cloud, revealing clear blue sky behind, demonstrates how temperature and moisture conditions have to be just right for cloud to form.

◄ SUNSET STRATIFORMIS

The evening light makes dramatic play of this altocumulus stratiformis formation, photographed in the skies above central England.

NEW ZEALAND PLATE STACK ➤

One of the peaks of New Zealand's Southern
Alps peeps into the bottom right edge of this
beautiful shot of altocumulus lenticularis. The
cloud has formed as the predominantly westerly
wind flows over the massive alpine barrier,
which runs almost the entire length of New
Zealand's South Island. Rising to over 3000m
(10,000ft) in many places, the mountains are
perfect for creating these pile-of-plates wave
clouds (see also Chapter Eight). Such air flows
also give some of the best conditions in the
world for glider pilots.

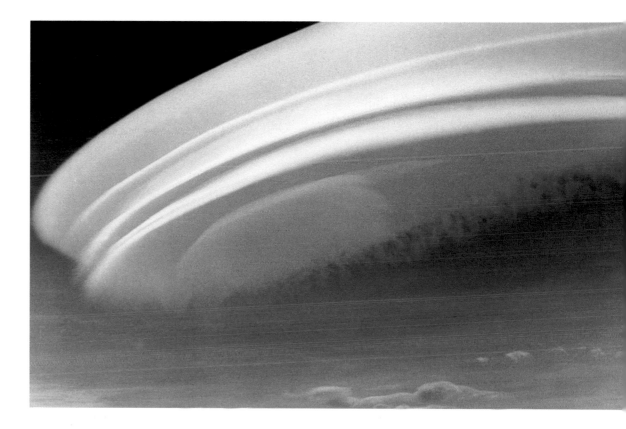

◄◄ "…WITH THUNDER LIKELY LATER"

These towers of medium cloud over the Channel Island of Jersey, Great Britain, are called altocumulus castellanus. Unlike the usual forms of altocumulus, these clouds are much taller than they are wide. Their name comes from the turreted look of the cloud tops, like castles in the air. They indicate instability in the atmosphere at the level of the cloud and often form prior to thunderstorms. If you see clouds like these, watch out for a storm later in the day.

◄ SWISS FLOCCUS

Lake Geneva, Switzerland, provides the setting for another lovely form of unstable medium-level clouds, altocumulus floccus. The name comes from that given to the tuft of hair at the end of a mammal's tail; hence this tufted form of altocumulus. These clouds have less well-defined edges than the classic form, being much more ragged; they also have greater height too. There is already a large cumulus cloud in the distance at the bottom of the picture, perhaps a sign of worse weather to come.

ALTOCUMULUS FEATHER ➤

This picture captures a beautiful feather duster of altocumulus. The cloud elements are quite small, aligned in a combination of horizontal and vertical stripes. Appearing as it does in a clear sky, the contrast between the white cloud and the blue is particularly marked. This is not commonly seen, but well worth looking for.

ALTOCUMULUS BANDS ➤➤

While the banded altocumulus in this shot shows the same basic form as classic examples of the cloud, what makes it look so different is a combination of the closeness of the individual cloud elements and the almost striped, or banded, appearance of the unusually dense cloud mass. This is due to some feature of the wind, temperature, and humidity within the cloud.

◄ HEATHROW SKY HIGHLIGHTED

This stunning example of a mackerel sky is dramatically enhanced by the late evening sunshine. Captured over London's Heathrow Airport, it even gives the cloud elements similar hues to those of the fish. Usually a mackerel sky means an altocumulus that has small, scalelike elements, but this one appears also to have the various shades of a real smoked mackerel.

ALTOCUMULUS FLOCCUS WITH TAILS ➤

These altocumulus floccus towers have become about as ragged as they get, their bases collapsing into virga, streaks of water droplets or ice crystals, falling out of the bottom of the cloud. Typically, virga fall vertically but they can curve away to one side as the droplets or crystals evaporate, shrink in size, and the wind then starts to carry them away.

FLOCCUS WITH RAYS ➤

This example of altocumulus floccus over the Isle of Wight, UK, shows just how wide a variety of forms these clouds can take. What makes them floccus is the puffy shape and the, albeit limited, vertical development. And the image also captures some striking crepuscular rays in the late evening sky. (See Chapter Eight.)

SUNSET ➤➤

Here is another example of altocumulus appearing in a classic mackerel-sky form, with some wonderful contrasting colours. These clouds really do create some beautiful skies, especially early or late in the day. Here they provide a memorable end to a summer's day in Bracknell, Berkshire, England.

"THEY ARE THIN AND DELICATE IN APPEARANCE, OFTEN RESEMBLING FEATHERS OR PLUMES"

CHAPTER SEVEN:
HIGH CLOUDS

High clouds are some of the prettiest we see. They are thin and delicate in appearance, often resembling feathers or plumes. This is because they occur high up in the atmosphere, above 6,000m (20,000ft). They are mainly made up of billions of tiny ice crystals alongside any super-cooled water droplets that have not found a particle upon which to freeze.

They take on three forms: isolated patches, known as cirrus; large sheets or layers, known as cirrostratus; and somewhat lumpy, though still very fine, patches called cirrocumulus. And what a variety of shapes and textures they all produce! From the hooks, barbs, and feathers of cirrus to the thin veils of cirrostratus, to the rippling seashore effect of cirrocumulus, these delicate high clouds really are fascinating in their beauty.

What's more, the ice crystals of cirrostratus clouds often produce some beautiful effects such as halos, arcs and coloured patches. Yet these are usually missed by most people as they go about their daily lives. If you see a thin veil of milky white cloud in the sky, take a closer look; you may see some unusual coloured patterns too. (More of these in Chapter Nine.)

CIRRUS

CIRRUS FIBRATUS AND A CONDENSATION TRAIL ➤

Cirrus clouds are generally the finest, most delicate high clouds of all. Being almost entirely ice-crystal clouds, they invariably have a very fibrous texture, as in this image taken late in the day near Cwm in Gwent, in Wales. The formal name for this cloud is cirrus fibratus, and as the name suggests, it is very fibrous – the classic appearance of all cirrus clouds. An aircraft condensation trail is starting to thin and distort in the high level winds; it is also casting a faint shadow on the cirrus to its left.

WISPY TWISTS ➤➤

The cirrus in this image has a wispy appearance because it is made up entirely of ice crystals. When these crystals become sufficiently large, they fall through the air creating a cloud with a significant vertical depth. The changing wind flow at these high levels often then causes the falling crystal trails to become twisted and irregularly curved, as we see here.

SPLENDID EVENING CIRRUS

This is yet another amazing formation of cirrus fibratus in the late evening sky, this time over Freshwater on the Isle of Wight, off the southern coast of England. Typical of the end of a fine summer's day, these wonderful clouds appear to radiate out from low in the sky and then to pulse in bands.

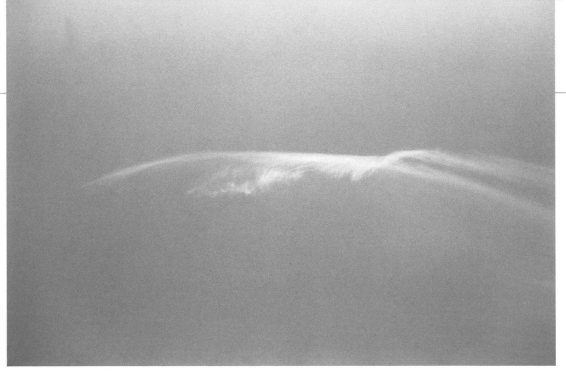

◄ CIRRUS FLYING HIGH!

What could be more spectacular than this single twisted filament of cirrus captured in Walcheren, in the Netherlands? Wonderfully white against the light blue summer sky, it appears to hang there like an enormous thin seagull.

◄ This unusual cloud is cirrus floccus, a high-cloud version of the floccus more usually seen at medium-cloud levels. Again we see the characteristic fluffy floccus nature of the individual cloud elements, despite the fact they contain a high proportion of ice crystals.

CLOUDS OR GHOSTS? ►

This amazing display was captured in an evening sky over Corrieshalloch Gorge, near Ullapool, in northwestern Scotland. Looking something like a gathering of a clan of ghosts, these twisted strands of fibrous cirrus cloud make quite a display. It is yet another example of the truly remarkable patterns created day by day in our skies by falling ice crystals twisting in the wind.

◄ TUFTS, HOOKS, AND FEATHERS

This is the epitome of a fine day: lots of clear blue sky; some beautiful cirrus cloud high in the sky; and a few puffs of small cumulus close to the ground. Taken near Padstow, Cornwall, England, the image provides an excellent example of tufted or hooked cirrus, which has the formal name of cirrus uncinus.

The shot below is another of the same delicate, almost feathery, tufts. Many of the filaments are even finer, but still have the same characteristic upward curved form at their extremities.

MAGNIFICENT GIANT FEATHER ►

This giant feather is an extraordinary display of cirrus spissatus (dense cirrus), and captures the best elements of the beauty of these clouds. Wispy tufts and filaments stream out around the edges, striations lie within the thicker main body of the cloud, and finally we see the glorious contrast between the fibrous white and grey cloud and the blue sky. Magnificent.

CIRROSTRATUS

RED SKY IN THE MORNING...

This vivid red sky is a perfect illustration of that old adage, 'Red sky in the morning, shepherd's warning.' The warning foretold rain later in the day. When you see a veil of cirrostratus like this at sunrise, it means that a warm frontal system is approaching. The cloud will slowly thicken into altostratus or nimbostratus, and lower during the day, almost certainly bringing rain by late afternoon.

◄ VEIL OF CIRROSTRATUS

Because cirrostratus is usually associated with frontal systems, particularly warm fronts, the cloud sheet is often seen spreading fully across the sky. In this image captured in Cove, in Hampshire, UK, the cirrostratus veil does just this. To trained observers, the extent of the cloud coverage makes the distinction allowing it to be correctly classified. Some aircraft condensation trails, straight white lines running across the lower left-hand side of the frame, and again towards the middle to the right-hand side, are also visible in this picture.

SUNSET WITH RAIN TO FOLLOW ➤

Cirrostratus rarely appears as a completely uniform, smooth sheet or veil; it is usually seen with some banding or changes in thickness or with fibrous patches within it. And it often appears with areas of cirrus cloud, too. This image from southern England contains almost all these characteristics, with one condensation trail appearing as a single wisp at the top left. The edge of this encroaching sheet of cirrostratus is somewhat ragged as it pushes into the clear, evening sky. Much of that edge is made up of cirrus, as are the thicker, whiter areas that appear beneath the main cloud veil.

◄ CIRROSTRATUS INVADING THE SKY

This is a typical example of cirrostratus, which, by definition, is said by weather observers to invade the sky. There is a wide fringe of beautiful fibrous filaments of cirrus followed by a sudden change to the thick white layer of cirrostratus. Low on the horizon, the cirrostratus has already thickened into a more bland layer of altostratus. The beauty of these clouds almost always belies the threat of worse weather to come.

PATCH AND BANDS ►

Observers would classify this area of cirrostratus as not increasing and not covering the whole sky. This lovely big patch is seen beside some banded cirrostratus and was pictured over the Isle of Wight, off the southern coast of England. This form of cirrostratus usually has a clear-cut edge, as in this case, though there is some other very thin cirriform cloud at the bottom of the picture.

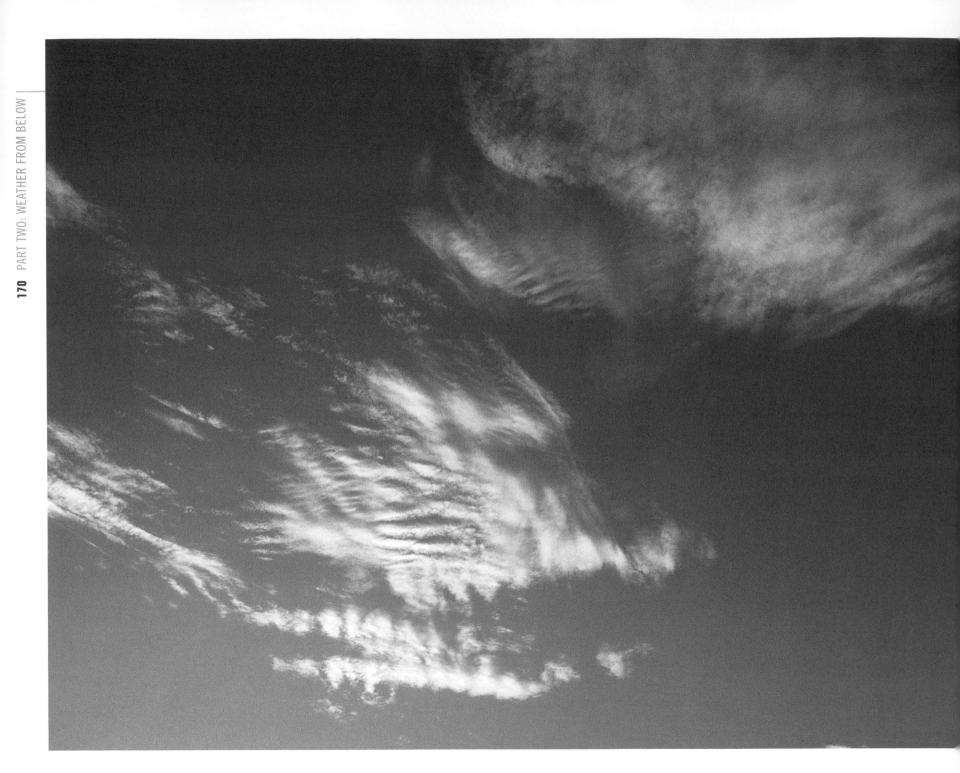

CIRROCUMULUS

◄◄ CIRROCUMULUS IN THE BLUE

The main characteristic of cirrocumulus cloud is its rippled appearance. True cirrocumulus cloud is not seen very often; it can easily be confused with the small cloud elements in certain forms of altocumulus. As with other high clouds, cirrocumulus is composed mainly of ice crystals, though some super-cooled water droplets may still be present. In this picture, there is a stunning contrast between the rippling clouds and the brilliant blue sky.

◄ MORE PATCH, LESS RIPPLE

This shot, captured in Finningley, Yorkshire, England, shows a larger area of cirrocumulus. It can often form within a patch of cirrus or a layer of cirrostratus. Either way, the expanse of cirrocumulus cloud can be quite extensive. Here, the elements are themselves patchier and lumpier than the more delicate ripples in the previous picture.

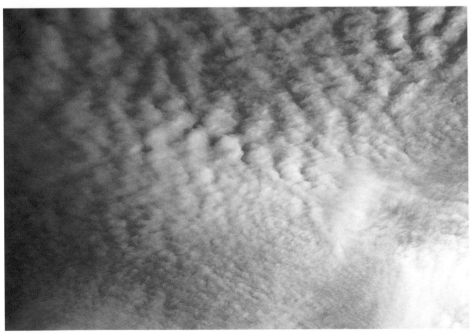

◄ CIRROCUMULUS IN RIPPLES AND PATCHES

Here, a layer of cirrostratus cloud has turned into cirrocumulus, typified by the tiny ripples in the cloud sheet. The elements on this occasion are virtually continuous, creating a rarely seen sheet of cirrocumulus. What's more, the cloud is quite thick, reducing the sun into nothing more than a bright patch (see also the photograph on the right).

In this image, from Finningley, Yorkshire, England, we can see a combination of patches and ripples. Notice how the individual elements vary in size from quite large and fluffy in the middle of the picture to very small, almost grainy, towards the bottom right.

CIRROCUMULUS OVER BIRMINGHAM ►

Taken at a slightly different time of day from the top-left picture, this image from Halesowen, Birmingham, England, has it all – blue sky and rippling and patchier cirrocumulus with some dense cirrus thrown in to blot out most of the sun. Mainly composed of tiny ice crystals, these high clouds can be thick enough to turn a warm summer's day into a noticeably cooler one.

A SHEET AND A PATCH ➤

Another pair of contrasting images, these show just how varied cirrocumulus can be. A late evening shot over Haslemere, in southern England, gives us an extensive rippling sheet, like the ripples in the sand near the water's edge. The lower image, caught in Llanbedrog, in Wales, appears as an almost isolated patch of wide, white ripples.

CIRROCUMULUS OVER KANSAS PLAINS ➤➤

This stunning example of tiny element cirrocumulus in a brilliant blue Kansas sky shows just how small the ripples can be. It may look like patches of cirrus cloud, some thick and some thin, but that is to miss the classic, though tiny, cirrocumulus structures in the cloud. The thicker areas of cloud, showing bright white, probably still contain super-cooled water droplets, while most of the rest of the clouds will be ice crystals.

"THERE ARE SOME CLOUDS THAT ARE VERY RARELY SEEN, AND SOME CLOUDS THAT DO NOT OCCUR NATURALLY"

CHAPTER EIGHT:
SPECIAL CLOUDS

In the world of weather observing, a wide range of clouds is seen everyday, though they take on a variety of forms and shapes. In addition to these, however, there are some clouds that are very rarely seen, and some clouds that do not occur naturally.

The most common of the man-made variety are those familiar white lines that often appear in blue skies, creating some beautiful patterns. These, of course, are condensation trails, or contrails for short. Such trails, looking somewhat like artificial cirrus clouds, form either from the exhaust gases of jet aircraft, or from the drop of pressure and the associated vortices, which occur at an aircraft's wingtip.

Later in this chapter we take a look at two naturally occurring clouds, which form in very particular circumstances. For the first, Kelvin Helmholtz billows, the atmosphere must be in a very precise state; any changes in stability and wind conditions and the remarkable patterns simply disappear. The second type of cloud, orographic, requires particular geography, mountains in fact, along with the right supply of moisture and wind flow. As you will see, these clouds can be really quite special.

CONTRAILS

◄ RED LINES IN THE SUNSET

Most of the cloud in the picture is either recent or old condensation trails, set in a vivid sunset sky with areas of real cirrus cloud. The recent ones are still in their narrow, straight lines, while the older ones have decayed into a range of shapes and features. We can see wide rippled bands of cloud across the middle of the picture, spreading out of the original contrail; towards the top, there is an even broader band of feathery ice-crystal cloud.

CONTRAILS OVER THE MIDLANDS ►

Set against a background of quite dense cirrus, this display of contrails, shot at Halesowen, in Birmingham, England, is full of features. The top contrail shows some vortices in the clear blue at the left-hand edge of the shot, while across much of the cirrus there appears to be a dark shadow of the contrail on the cirrus cloud itself. The lower trail shows vortices along much of its length, except where it is hidden by a patch of cirrus that is slightly lower in the sky. Contrails often follow the same directions because aircraft fly along designated routes called airways, somewhat like motorways in the sky.

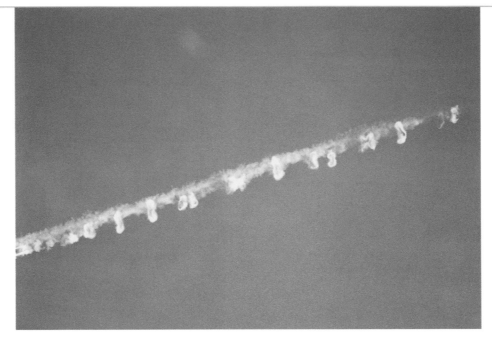

◄ LOOPS AND LINES

An aircraft creates considerable turbulence as it passes through the atmosphere, and a contrail marks where an aircraft has been. Some of that turbulence is transferred into small swirls or vortices in the contrail cloud; these are characterized by the loops and figures of eight we see. This photograph of a single contrail was taken over Benson, in Oxfordshire, UK. It is already breaking down into beautiful little cloudlets, called vortex loops.

◄ Another series of contrails running across a patch of thin cirrus in an otherwise brilliant blue sky in Derbyshire, England.

CIRRUS WITH CONTRAILS ➤

Here, most of the contrails that streak across a cirrus sky are pretty much intact but others are well on their way to dispersing. There are some falling streaks of cirrus ice crystals in the top right-hand corner. Below this, the contrail is casting a marked shadow on the cloud beneath. At the bottom of the picture there is a criss-cross of trails running across the housetops while at the top left-hand corner, one old contrail is breaking up into small patches of cloud, not unlike the cirrus beside it.

◄ SMOKE RINGS IN THE SKY

This sweeping ring of cloud is probably the condensation trail left by an aircraft as it made a completely circular flight. Patterns like this are often made by routine surveillance aircraft or by military aircraft on manoeuvres. Such rings are more common than you might imagine, especially in those parts of the world where military aircraft routinely operate. This one was photographed in Durham in the north of England.

CONTRAIL STREAKS AND PUFFS ►

Originally born as a single, straight aircraft contrail, it sometimes doesn't take long before they change shape and form. This one is already breaking up into a series of streaks and puffs, and will probably disperse completely before too long.

KELVIN HELMHOLTZ

KELVIN HELMHOLTZ BILLOWS

Normally, wave clouds form when the atmosphere is stable and the wind speed and direction are steady. If the atmosphere becomes unstable and the wind flow rapidly changes in speed and direction — wind shear — then these remarkable, tumbling, twisting clouds can form. They are known as Kelvin Helmholtz clouds, or billows, taking their name from the two scientists who developed the theory of fluid flow behind their formation. They are very unusual, and whether seen in full daylight as at Laramie, Wyoming, USA, or in the coloured evening sky of Farnham, Surrey, England, they provide very beautiful cloudscapes.

OROGRAPHIC CLOUDS

◄ WAVE CLOUDS ACROSS THE WORLD

Is it a cloud? Is it a flying saucer? Mount Ruapehu, New Zealand, has produced a spectacular 'pile-of-plates' wave cloud in this shot. Created by wind flow over the mountain, this example rises right up into the realms of high clouds, as can be seen from the cap of fuzzy ice-crystal cirrus cloud above the main wave-cloud layers. The process that forms these wonderful clouds is called orographic lifting, the term for air rising over a mountain barrier.

This wave cloud was photographed at Loch an t-Sailen on the Isle of Islay, Scotland. As with the picture above, these remarkably elegant wave clouds remain stationary, linked to the high ground barrier that creates them, hence their name, standing wave.

MATTERHORN BANNER CLOUD ➤

This is a splendid example of a banner cloud streaming like a flag off the peak of the Matterhorn in Switzerland. The physics of banner clouds is not well understood. The basis seems to be that pressure falls on the lee side of the peak, away from the prevailing wind. This induces air to rise up the slope of the mountain, and if this air is moist, then the moisture will cool and condense to create the banner blowing in the wind.

◄ MERINGUE ON A CAKE PLATE

This is a stunning example of orographic lifting over Mount Ruapehu in New Zealand. It soars from the very top of the mountain, right up into the ice-crystal domain of high clouds. It gives the impression of a brilliant blue piece of china, topped off with fluffy white meringue.

◄ STANDING WAVE OVER SCAMPTON

This form of altocumulus lenticularis (bottom left), photographed over Scampton, in Lincolnshire, England, is known as a standing wave cloud. The wave part is created when wind flowing over hills sets the air into a wave-like motion; the standing means the cloud remains in the same place, appearing to stand still while the moving air passes through it.

NOCTILUCENT CLOUD

LINCOLNSHIRE NOCTILUCENT CLOUD ➤

If you ever see clouds like these, you have enjoyed a rare experience. They are not actually true clouds. The phenomenon is produced by the reflection of sunlight from masses of volcanic or meteoric dust particles about 50 miles up into the atmosphere. Noctilucent clouds can only be seen between latitudes 45° and 60° north or south, from mid-May to August. They appear in a clear sky about 15 minutes after sunset and are seldom seen at an altitude of more than 10°.

"CREPUSCULAR RAYS APPEAR AS GOLDEN SHAFTS OF LIGHT AROUND THE EDGES OF CLOUDS, USUALLY WHEN THE SUN AND CLOUDS ARE LOW IN THE SKY"

CHAPTER NINE:
OPTICAL PHENOMENA

As we have seen, the world of weather creates many different types of remarkable effects. Some of the most beautiful are created directly by sunlight. For example, crepuscular rays appear as golden shafts of light around the edges of clouds, usually when the Sun and clouds are low in the sky.

When nature causes the light to break down into its seven parts, the primary colours, the effects can be even more striking. Light might be reflected through raindrops, producing the familiar rainbow. Usually there is one rainbow on its own, but sometimes there is a pair or even as many as three. Higher up in the sky, light refracted, or bent, through ice crystals in the thin high cloud cirrostratus (see Chapter Two) produces a whole range of mainly circular or curved patterns, including halos, arcs, cloud pillars, and mock suns.

There are a few other optical phenomena, including coronae, glory, and mirages, to name but three, which are much more unusual. And although not a weather-related phenomena at all, because it is linked to electromagnetism, the aurora borealis creates fantastic shimmering lights in the northern night sky, simply too stunning to be ignored.

ARCS

◄ CIRCUMZENITHAL ARC

It is the high ice-crystal clouds, particularly cirrostratus, that create some of the most spectacular optical phenomena, and the circumzenithal arc is one of the most colourful. It is usually comparatively short, covering just a small part of the sky, but can be very vivid in its colouring. A circumzenithal arc occurs in the same direction as the Sun but quite a lot higher in the sky. As the name suggests, it has a curved shape with the inside of the curve towards the Sun.

COLOURED TUFTS ►

The circumzenithal arc display in this image is not so bright but still coloured, and this is more typical than the display opposite. Sadly these features usually do not last long. They require the cloud, in this case tufted cirrus, to be in a precise position in relation to the sun, and the crystals within the cloud to be of certain shape and size. Watch out for a hazy, milky sky of cirrostratus cloud and you might just be lucky enough to see an arc like this.

CORONAE

◀◀ CORONA AROUND THE MOON

It is most common to see a corona around the
moon at night. They do occur in daylight around
the sun but the intensity of colour is too faint to
be seen. In this image, there is a reddish ring
towards the edges of the cloud, often all that
can be seen. However, on rare occasions, it is
possible to see the full range of rainbow colours
working outwards from violet to red.

◀ LUNAR CORONA OVER SKYE

The size of a corona can vary considerably, the
determining factor being the size of the cloud
particles that produce it. Small particles create
a large corona, while large particles create a
small one. Coronae are usually formed when
a thin layer of water-droplet cloud obscures
the moon. In this case, captured on the Isle of
Skye, Scotland, we see a wide corona formed
in an area of altocumulus. The texture of the
cloud and the uniformity of the size of its water
droplets both affect the colouring of the corona
that is being produced.

PARHELIA/ MOCK SUNS

◄ MOCK SUN AT SEA

Mock suns, technically known as parhelia, are known as sun dogs in America. Whatever their name, they are bright arcs or brilliant spots seen some distance from one or both sides of the Sun and at the same level in the sky. This photograph, taken at sea, shows the partially obscured Sun near the left edge of the frame with a bright mock sun and a coloured arc of a halo just emerging from behind a large cumulus cloud (right). Both the mock sun and the halo arc have been created by light play in the high cirriform ice crystal clouds.

MOCK SUNS, COLOURED AND PLAIN ➤

(Top Right) This small bright mock sun was captured in Brynmawr, in South Wales, and is typical of the coloured-spot form of this phenomenon. Once again, the sun is hidden behind a cumulus cloud, and the mock sun has formed in a layer of cirrostratus.

(Bottom Right) This yellow evening sky has a bright white mock sun superimposed on it well to the right of the Sun, which appears at the left-hand edge of the picture. The result is a yellow mock sun.

IRISATION

◄◄ IRISATION ON BLUE

Irisation, the random colouring of parts of
clouds, is another unusual optical phenomenon.
It usually occurs at the edges of cirrus,
cirrocumulus, and altocumulus, and may be
white, green, or red. In this picture, the main
colouring is white, immediately above and
around the Sun, but above that, there is an
area of pink and green irisation in the iridescent
cirrocumulus ripples.

◄ IRISATION ON YELLOW

The overall yellow colouring in this image is
amazing in its own right but there is some extra
irisation too. The Sun is setting and is hidden
behind the band of cloud close to the horizon.
Above its position, there is a brilliant white edge
to an area of altocumulus. If you let your eye
drift over the patch of blue sky and into the
edge of the upper area of cloud, there are some
tinges of red and green – irisation.

SUN PILLARS

SUN PILLAR ➤

A sun pillar is (usually) a very bright white column extending up from the setting or rising Sun, and at sunset they can become red. Whatever the colour, they display a marked glittering effect. This is because they are caused by the direct reflection of sunlight off the surface of the tiny ice crystals in cirriform clouds, each crystal behaving like a minute moving mirror. This example from Dishforth, in Yorkshire, England, shows the classic white column in all its splendour.

SUN PILLAR SURPRISE ➤➤

This sun pillar in southern England looks almost like an exclamation mark in the sky. You can see a bright column, typical of a sun pillar, but at its base there is a brighter patch in the sky, which is probably caused by a thicker patch of cloud with a greater concentration of ice crystals. The colouring of the sun pillar is dominated by the overall effect of the setting Sun.

RAINBOWS

◄ CLASSIC PRIMARY RAINBOW

Rainbows are formed when sunlight is bent
and reflected by raindrops. This typically takes
place on a day of sunshine and showers, when
the Sun is out in one part of the sky, behind
the observer, but a shower of rain is occurring
in another part. The bending, or refracting, of
light means it is split into the seven primary
colours (red, orange, yellow, green, blue, indigo,
and violet) with red on the outside of the bow
and violet inside. The brilliance of the colours
is down to the size of the raindrops, with large
drops forming the most brightly coloured bows.

TWO RAINBOWS – PRIMARY AND SECONDARY ►

When the sunlight striking the raindrops is
reflected twice inside each drop, a slightly
fainter secondary bow forms outside the primary
bow, as in this example from Embleton Bay,
Northumberland, England. The colours of the
primary bow run in the usual sequence with
red on the outside; those in the secondary bow
occur in the reverse order, with the red arcs of
each bow facing each other.

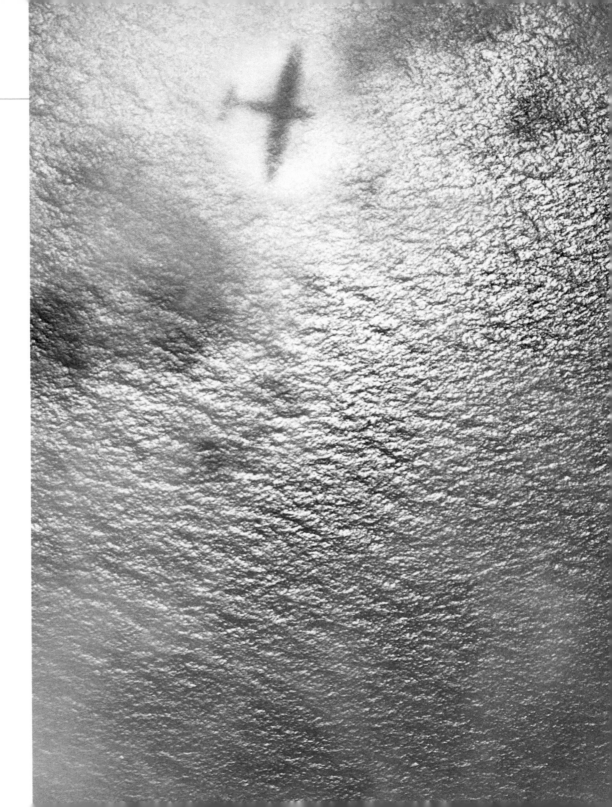

GLORY

GLORY OVER THE SEA ➤

This phenomenon is commonly called glory, but it is also known as the spectre of the Brocken after the mountain range where it is seen relatively often. The glory, a series of coloured rings around the shadow of an object, occurs in two situations. It can surround an aircraft's shadow cast on a layer of cloud below or encircle a mountain climber's shadow cast on fog or cloud in a valley. In this shot from the Whitsunday Islands, Australia, the aircraft's shadow is seen on some very thin stratus cloud almost at sea level.

GLORY ON CLOUD ➤➤

In this image, again taken from an aircraft, the glory is visible on a thick layer of cloud. The shadow of the plane is less visible but the coloured rings are very much in evidence. As with coronae, the process creating the coloured rings is the spreading and refracting of sunlight by cloud droplets.

CREPUSCULAR RAYS

◄ CUMULUS CREPUSCULAR RAYS

Crepuscular rays, or sunbeams, need both the Sun and clouds to be in the right place at the right time. They are always spectacular to see, and occur in three ways. This image caught in Dundee, Scotland, is the one most seen, when the Sun is hidden behind a large cloud and light rays stream from round its edge. The effect is usually bright rays of light with darker bands between them. Other forms are seen when sunlight beams down through a hole in a cloud or upwards from the edges of cloud at sunrise or sunset.

SUPER TROOPERS IN THE SKY ►

This is a shot of classic crepuscular rays. The light rays, somewhat like spotlight beams, blaze up from the edge of the cumulus cloud, itself starting to catch the yellow of the Sun. It is easy to see how the jagged outline of this cloud affects the light and dark patterns of the rays, always an amazing weather effect to see.

◄◄ WHITE GOLD AT PORT ERIN

Shot at Port Erin on the Isle of Man, this glorious golden image has crepuscular rays streaming from behind some clouds and cutting across others. There is a real burst of white rays immediately above the sun, and the top of the large cumulus cloud (top left) is catching some of the last real daylight too. In the lower left quadrant, clouds are already slipping into the darkening evening twilight.

◄ LIGHT AND DARK RAYS

(Top) This sea and cloudscape in blue is finished off by the bluish white rays bursting out from behind the cumulus cloud off the Welsh coast at Aberystwyth.

(Below) In this shot of a Jamaican sunrise, it is the shadows produced by irregular parts of the cloud that break into the brightening sky and give it a reverse ray effect – dark out of light. The red tinges from the just-rising Sun simply add to the show.

CLOUD
ILLUMINATION

At dawn and dusk, we see clouds mainly as shades of red. This is because light at the red end of the spectrum penetrates the atmosphere better, while the light at the blue end of the spectrum is more scattered before reaching the cloud, and is therefore less visible.

◄◄ RED, WHITE, AND BLUE

Perhaps the best place to capture sunsets anywhere in the world is on a western coastline, where there is nothing to spoil the view. This superb red sky was captured at Square and Compass in Pembrokeshire, not far from the coast in western Wales. There is an area of low cloud, probably stratocumulus, towards the horizon, while many of the other clouds are areas of cirrus at different heights. There are even the remnants of a contrail still catching the last rays of white light before the sun finally sets.

◄ WELSH GOLD

When the cloud cover is split or layered, the setting Sun can burst through the gaps making a dramatic contrast between the dark clouds and the dazzling evening sunlight. This pair of Welsh images, both taken near Ebbw Vale, south Wales, provide just such a display.

◄ PATHWAY TO HEAVEN

Sunset over the sea always adds another dimension to the scene, offering us a pathway to heaven in the form of a golden strip of light that runs towards us across the sea. With stratocumulus and altocumulus clouds at different levels catching the light in different ways, and the sound of gently lapping waves, the pathway to heaven scenario is always tranquil and comforting.

TREES IN THE EVENING LIGHT ➤

These partially lit conifers stand out starkly against the fading cumulus clouds low on the horizon in this evening sky. In the middle of the picture, thin strands of darkening cirrus can be seen while the main cloud features are the golden undersides of bulbous cumulus.

GOLD ON BLUE ➤

This lovely composition presents a beautiful sunset over the River Exe, in Lympstone, Devon, England. The upper sky is mainly clear and blue with some brilliant white patches of cirrus, high enough still to be lit by the full light of the Sun. Down low, almost on the horizon, the setting Sun is tinting all the clouds with shades of red, yellow, and orange. This effect is caused by the scattering and absorption of light of all the other wavelengths, leaving only the glorious golden colours of a sunset.

◄ BRIGHT NEW DAY

In this picture of sunrise, taken in Berkshire, England, a large area of altocumulus is lit from underneath by the rising Sun. This is an example of white sunlight creating the spectacle, rather than the typical golden shades. The pretty cloud pattern also plays a significant part in this display.

MIRAGE

ANTARCTIC MIRAGE ➤

The term mirage is used when strangely altered or duplicated images occur on the horizon at ground level. They are caused by the bending (refraction) of light due to significant changes in the density of the air close to the ground. In this image, taken over the Ross Sea in Antarctica, the mirage takes the form of a distorted horizon of ice cliffs, which seem to have reflections above them.

MIRAGE OF TWO SUNS ➤➤

What is real and what is imaginary? – that is the paradox of a mirage. In this sunset scene, the mirage is quite obviously a complete duplication of just part of the atmosphere, and so a second segment of the Sun appears out of the sea. Which is the true horizon? The eye and brain can find it hard to analyze exactly what they see.

AURORA

◄ AURORA RAYS

An aurora, or more formally, an aurora borealis when it occurs in the northern hemisphere, has no connection with the weather. It is an electrical phenomenon usually associated with magnetic storms above 80km (50 miles), in the highest reaches of the atmosphere. Aurora borealis takes its name from Aurora, the Roman goddess of the dawn, and Boreas, the Greek name for the north wind. Auroras produce many wonderful, slightly eerie, coloured displays. These two pictures are of rays, the top one glowing red and extended into a curtain, the lower aurora brilliant blue-green rays, looking much like searchlights.

AURORA CURTAIN ➤

Aurora that occur in northern latitudes of the northern hemisphere are known as Northern Lights. The displays are not static, and they frequently change colour, moving and rippling in the night sky. This example is of brilliant bluish-white curtains. The white spots in the images are stars, and can be used to identify the part of the sky in which the displays occur, in this case the constellation of Leo. The northern lights are always a truly amazing and unforgettable sight.

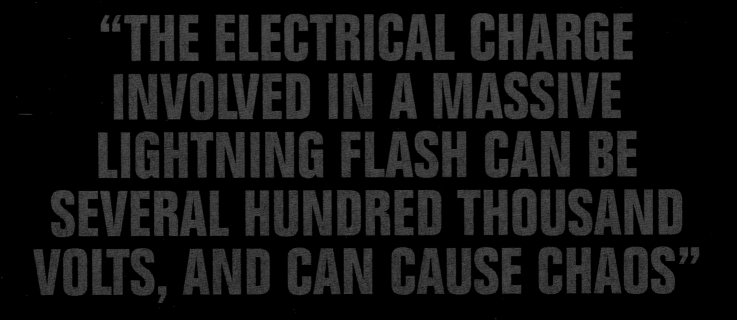

"THE ELECTRICAL CHARGE INVOLVED IN A MASSIVE LIGHTNING FLASH CAN BE SEVERAL HUNDRED THOUSAND VOLTS, AND CAN CAUSE CHAOS"

CHAPTER TEN:
EXTREME WEATHER

Extreme weather can come on a number of scales and intensities. In this chapter, we take a look at local extreme-weather phenomena. These are the events that may affect just one town or city, part of a town, or one part of an ocean. They may be localized in their extent, but their effects may have devastating consequences.

First we focus on lightning. Created mainly by mature cumulonimbus clouds, lightning is amazingly beautiful, yet frighteningly powerful. The electrical charge involved in a massive lightning flash can be several hundred thousand volts, and can cause chaos.

Then we turn to tornadoes, those twisting areas of intense low pressure that move erratically across the land, wreaking havoc wherever they go. There is something fascinating yet sinister about these twisters. Perhaps it is simply their awesome local power; so fast, so fierce, and sometimes, so final.

Lastly we move offshore to examine, using Admiral Beaufort's original wind scale, how extreme wind can affect the state of the sea. Building up from ripples to massive walls of water, we see how waves grow as the wind increases.

LIGHTNING

◄ SPIDER LIGHTNING

Lightning is the most common extreme-weather event, and is associated with just one type of cloud, cumulonimbus (see Chapter Five). Normal lightning occurs when an electrical charge builds up between the top and the bottom of a cumulonimbus cloud and the ground beneath, and is then discharged. Significant variations in charge inside a cloud can lead to multiple internal flashes, sometimes identified as spider lightning, as in this dramatic image, from Woodward, Oklahoma, USA.

VALENTINE ZAPPER ►

In large cumulonimbus clouds, a huge buildup of charge occurs between the cloud and the surface of the ground. At a critical level of charge, we see a brilliant, single, momentary cloud-to-ground flash or stroke. Often lightning flashes come straight down; sometimes they curve or fork; and sometimes they cut right across the sky. This lightning flash, captured in Valentine, Nebraska, USA, has ripped out of the cloud and zapped diagonally down to Earth.

◄◄ FORKED LIGHTNING

This shot looking down onto the city of Budapest, Hungary, catches an interestingly complex lightning flash. Probably experienced as a single amazing lightning flash, the camera identifies one main stroke with several lesser strokes, all making their way to ground. As the electrical charge builds up in the cloud and on the ground, various paths of least resistance are created in the atmosphere. That is what leads to multiple or forked flashes rather than the single stroke.

◄ KANSAS CRACKER

This multiple-stroke image from Sharon Springs, in Kansas, has one flash racing across the base of the cloud while another hurtles down to the ground, forking as it goes. Significantly more than half of all lightning flashes, or discharges, occur within thunderstorm clouds, yet it is the cloud-to-ground flashes we most readily recognize.

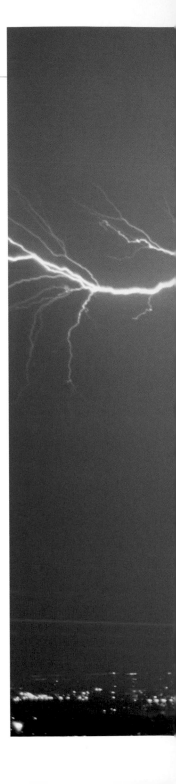

◄ LIGHTNING, UK TO USA

Taken using time-lapse photography, this picture has captured three separate lightning flashes over a few seconds. But this is no giant American storm. In fact, it was caught in Bracknell, Berkshire, for many years home to the UK's Met Office.

◄ In this Arkansas super-cell powerhouse, you can almost hear the lightning crackle. There is sheet lightning from a discharge within the cloud and at least two strokes from the base of the cloud to the ground.

MASSIVE LIGHTNING FLASH ➤

Sometimes lightning can discharge from a cloud into apparently clear air, and there it fades. This spectacular image from Budapest, in Hungary, displays several fading forks discharging in the air beside one massive brilliant white stroke that hammers right into the city. It probably meant problems for some residents that night, and not just a noisy, sleepless night.

TORNADOES

◄◄ AUSTRALIAN TORNADOES

This intriguing pair of photographs captures the same tornado at different times as it moved on its way in Northam, western Australia. The terrain is very dry and dusty, and the red dust has been sucked up into the tornado column. This one is really quite small, with a tight circulation footprint. More typically, tornadoes are a few hundred metres across.

◄ DUST DEVIL IN COLORADO

Right at the bottom of the scale of tornado-type phenomena is the dust devil, such as the example captured here in Walsenburg, Colorado, USA. Tornadoes form in the downdraughts from very powerful thunderstorm clouds; dust devils form in updraughts from local heating in hot, dry, sunny conditions. This beautiful wispy example appears delicate and tenuous against the brilliant blue sky.

NEBRASKA MENACE

This menacing black mass combines a super-cell thunderstorm with a tornado. Tornadoes occur on every continent but are most commonly reported in the USA, where there are, on average, 1,000 each year. Most of these occur in the southeastern states and the Great Plains – an area sometimes known as Tornado Alley, where this one was born.

◄◄ RARE TWIN TORNADOES

Tornadoes are very intense areas of low pressure accompanied by very strong winds. One tornado can cause dramatic damage, so the concept of two tornadoes in the same place at the same time goes way beyond normal experience. It is a truly rare event, but not unique. These two, born out the same super-cell storm cloud, were caught in Nebraska, the birthplace of many tornadoes. They both show the typical funnel cloud extending down from the base of the storm.

◄ FUNNEL CLOUD

When a funnel cloud grows out of the bottom of a giant cumulonimbus thunderstorm cloud, it is very likely a tornado will follow. The funnel cloud marks the column of rapidly rotating air at the base of the main storm cloud; if the funnel comes right down to ground level, then it is said to have touched down and becomes a tornado. This one appears to be still up in the air but a hazy rotating column of air can be seen beneath it.

◄ TIGHT DUSTY TORNADO

This small Nebraska tornado has a very tight, narrow vortex. Small tornadoes may only be 50–100m (165–330ft) across with wind speeds of just 65km/h (40mph). But the biggest can be many hundreds of metres in diameter and reach wind speeds of 485km/h (300mph), when significant damage is inevitable in built-up urban areas.

FARM BUILDINGS IN PERIL ➤

As the funnel cloud drops out of its parent super-cell cumulonimbus cloud, this farm seems to be right in the path of the forming tornado. The immense power focused in tornado vortices means they easily pick up animals and vehicles, and destroy buildings. Tornado winds can even drive pieces of debris right through walls and destroy apparently safe places, the kind of devastating effects we have all seen on television.

TWISTING TORNADO PIPELINE

Commonly called twisters, tornadoes really are some of the most dramatic weather events in the world. The funnel cloud and vortex of this tornado is behaving in an extraordinary way, beginning as a long, almost horizontal column and then twisting almost 90° before touching down. Captured in Clay County, Nebraska, this unusual image clearly shows tornado events can be as highly individual as they are deadly.

STORM-FORCE WINDS

SEA STATES

Rear-Admiral Sir Francis Beaufort (1774–1857) created the Beaufort Wind Scale. It has two main sets of criteria: those for observers on land, and sea states for observers at sea. The latter formed the original basis for wind observations. There are 12 wind forces and sea states, from Force 0, or calm, to Force 12, hurricane force, usually abbreviated to F0, F1, and so on, up to F12.

This series of pictures takes us from the quiet state of Calm (top left) to Force 5 (bottom right).

We can follow the build from ripples (F1) up to

Calm	F1	F2
F3	F4	F5

small wavelets (F2) through large wavelets (F3) to small waves (F4) and moderate waves (F5). Little white caps appear at F4 and there are very pronounced white horses by F5.

This is where things start getting interesting. This second series of pictures takes us from Force 6 up to Force 11.

F6	F7	F8
F9	F10	F11

We start off at large waves with white foam crests (F6) to breaking waves blown in streaks (F7) and moderately high waves of greater length (F8), known as fresh gale. By now, ships are really being tossed about. We then get high waves, dense streaks of foam, and toppling crests, but strangely, seas appear to flatten (F9), then very high waves appear with overhanging crests, the sea taking on a white appearance (F10). At Force 11, we experience exceptionally high waves behind which small- to medium-sized ships may disappear from view. This is not a great time to be at sea.

◄ HURRICANE FORCE 12

This shot typifies what it is like in hurricane Force 12 winds at sea. There are massive waves, stretching to 9–12m (30–40ft) from trough to crest, with long distances between successive crests. The rise and fall of a vessel is dramatic, and when a ship is heading into such waves, huge amounts of water break over the bows and wash over the deck. White water and spray seem to be everywhere.

HURRICANE FORCE 12 ➤

Here is another Force 12 picture. This time the ship's bows are digging into an oncoming wave. These waves are daunting walls of water, yet for the seasoned mariner, encountering such waves is all in a challenging day's work. But waves may also be coming from more than one direction, and this is what can give a vessel the most uncomfortable motion of all. Not only does it rise and fall on the main waves, but it also rolls from side to side in the cross-waves.

INDEX

PICTURE CREDITS

Front cover: Roger Coulam
Back cover: Jacques Descloitres, MODIS Land Rapid
Response Team
Page 6 and page 110: Roger Coulam

PART 1

Dundee University 22, 23, 24, 25, 26, 27
GeoEye and NASA SeaWiFs Project 32, 38, 39, 40
NASA 18, 19
NASA/Jesse Allen, Earth Observatory 71
NASA/Jesse Allen, Earth Observatory, using data provided courtesy of NASA/GSFC/METI/ERSDAC/JAROS, and the US/ Japan ASTER Science Team 99
NASA/Jesse Allen, using data obtained from the Goddard Earth Sciences DAAC 53
NASA/Aqua 72, 73
NASA/GSFC/Jesse Allen, based on data from the MODIS Rapid Response Team 50
NASA/GSFC/Bob Cahalan 50
NASA/GSFC/Jacques Descloitres, MODIS Land Rapid Response Team 28, 29, 30, 37, 44, 46, 47, 48, 49, 55, 56, 58, 59, 62, 64, 65, 68, 69, 77, 78, 86, 87, 88, 89
NASA/GSFC/JPL, MISR Team 61
NASA/GSFC/LaRC/JPL, MISR Team 45, 74, 75
NASA/GSFC/MITI/ERSDAC/JAROS and US/Japan ASTER Science Team 96
NASA/GSFC, MODIS Rapid Response Team 35, 63, 94, 95
NASA/GSFC/Jeff Schmaltz, MODIS Rapid Response Team 34, 52, 54, 57, 70, 79, 92, 100, 101
NASA/GSFC/Scientific Visualization Studio 80
NASA/GSRC/Reto Stöckli (land surface, shallow water, clouds). Enhancements by Robert Simmon (ocean color, compositing, 3D globes, animation). Data and technical support: MODIS Land Group; MODIS Science Data Support Team; MODIS Atmosphere Group; MODIS Ocean Group Additional data: USGS EROS Data Center (topography); USGS Terrestrial Remote Sensing Flagstaff Field Center (Antarctica);

Defense Meteorological Satellite Program (city lights) 16, 17, 20, 21
NASA/ISS Crew Earth Observations experiment and the Image Science & Analysis Laboratory, JSC 84
NASA/JSC, Earth Sciences and Image Analysis Laboratory 51, 76
NASA/Robert Simmon, based on Landsat-7 data from the Global Land Cover Facility 98
NASA/Robert Simmon, NASA's Earth Observatory, based on data copyright Space Imaging 97
NASA/Space Science and Engineering Center, University of Wisconsin 36
NASA/Reto Stöckli, with the help of Alan Nelson, under the leadership of Fritz Hasler 81
NASA/USGS Eros Data Center, based on data provided by the Landsat Science Team 90, 91
NASA/USGS Eros Data Center, Satellite Systems Branch (part of the Landsat Earth as Art series) 93

PART 2

D. G. Allen 225
Prof. A. H. Aver Jnr 184
A. Best 210
J. Bowskill 140 top, 186 top, 188 top
C. S. Broomfield 133, 136 bottom, 138 top, 149, 160 bottom
D. Brown 214
D. M. G. Buchanan 150
Stephen Burt 117, 122, 132 top, 147, 148, 153, 167, 180 bottom, 198, 199, 202, 205
Jane Corey 124, 156, 157, 163, 174 bottom, 192, 214 top, 216 top and bottom, 218, 219
Roger Coulam 106, 107, 108, 109, 110, 111, 112, 113, 114, 115, 116, 118, 120, 121, 130, 175, 194, 195, 196, 207, 208, 228, 229, 230, 231, 232 bottom, 233, 235, 237, 238, 239, 240, 241, 242
Crown copyright 140 bottom, 151, 160 top, 171, 172 bottom, 181, 185, 188 bottom, 244 left: top and bottom, right: bottom, 247
W. M. Cunnington 209
C. Dake 143

Marion Foreman 123
Jim Galvin 127, 132 bottom, 136 top, 137, 187 bottom, 211
Mrs A. Gilkes 186 bottom
J. Haines 234 right
P. D. Harris 143
Dr R. Harris 182
M. I. Holmes 162 bottom
R. Howes 145, 146
R. N. Hughes 182
S. Jebson 221
A. McClure 144
D. McConnell 144
A. McHardy 212
R. W. Mason 166, 220
P. J. May 234 left
J. R. S. Morgan Grenville 208
P. J. B Nye 129, 180 top, 193, 204
W. G. Pendleton 197, 201 bottom
W. Pike 164
R. K. Pilsbury 126, 138 bottom, 139, 142, 152, 159, 168, 169, 170, 203, 215 bottom, 223, 232 top,
J. M. Pottie 161
C. G. Roberts/Crown copyright 128, 141, 172 top, 173, 178, 179
J. Shepherd 249
K. B. Shone 213
D. A. R. Simmons 224 top and bottom, 225
J. C. Smart 200, 201 top
J. P. Thomas 174 top
J. Thomson 248
J. Walton 162 top
M. J. Wood 222

Every reasonable effort has been made to contact the copyright holders of images. If there are any omissions or inaccuracies David & Charles will be happy to corrrect this at a subsequent printing.

ACKNOWLEDGMENTS

As well as expressing my thanks to the editorial team at David & Charles, I would like to acknowledge the help of a few more people.

Firstly, Andy Yeatman for 'introducing' me to this project; thanks Andy! Secondly to two good friends and ex-colleagues – John Charlesworth and Pete Trevelyan – for their confirmation of the facts regarding two images; and thirdly to my brother, Alan Higgins, for clarifying one piece of obscure spelling.

Then finally, yet underpinning the whole project, I most warmly acknowledge the patience and support of my precious wife Sue. Thank you!

GH